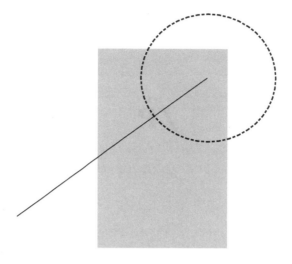

THE FREE PRESS   New York London Toronto Sydney Singapore

EUCLID'S WINDOW

THE STORY OF GEOMETRY
FROM PARALLEL LINES
TO HYPERSPACE

LEONARD MLODINOW

THE FREE PRESS
A Division of Simon & Schuster, Inc.
1230 Avenue of the Americas
New York, NY 10020

Designed by Jeanette Olender
Illustrations by Steve Arcella

Manufactured in the United States of America

1 3 5 7 9 10 8 6 4 2

Library of Congress Cataloging-in-Publication Data
Mlodinow, Leonard
Euclid's window : the story of geometry from parallel lines
to hyperspace / Leonard Mlodinow.
p. cm.
Includes bibliographical references and index.
1. Geometry—History. I. Title.
QA443.5 .M56 2001
516'.009—dc21   00-54351

ISBN 0-684-86523-8

To

Alexei and Nicolai, Simon and Irene

CONTENTS

INTRODUCTION

Twenty-four centuries ago, a Greek man stood at the sea's edge watching ships disappear in the distance. Aristotle must have passed much time there, quietly observing many vessels, for eventually he was struck by a peculiar thought. All ships seemed to vanish hull first, then masts and sails. He wondered, how could that be? On a flat earth, ships should dwindle evenly until they disappear as a tiny featureless dot. That the masts and sails vanish first, Aristotle saw in a flash of genius, is a sign that the earth is curved. To observe the large-scale structure of our planet, Aristotle had looked through the window of geometry.

Today we explore space as millennia ago we explored the earth. A few people have traveled to the moon. Unmanned ships have ventured to the outer reaches of the solar system. It is feasible that within this millennium we will reach the nearest star—a journey of about fifty years at the probably-some-day-attainable speed of one-tenth the speed of light. But measured even in multiples of the distance to Alpha Centauri, the outer reaches of the universe are several billion measuring sticks away. It is unlikely that we will ever be able to watch a vessel approach the horizon of space as Aristotle did on earth. Yet we have discerned much about the nature and structure of the universe as Aristotle did, by observing, employing logic, and staring blankly into space an awful lot. Over the centuries, genius and geometry have helped us gaze beyond our horizons. What can you prove about space? How do you know where you are? Can space be curved? How many dimensions are there? How does geometry explain the natural order and unity of the cosmos? These are the questions behind the five geometric revolutions of world history.

It started with a little scheme hatched by Pythagoras: to employ mathematics as the abstract system of rules that can model the physical universe. Then came a concept of space removed from the ground we trod upon, or the water we swam through. It was the birth of abstraction and proof. Soon the Greeks seemed to be able to find geometric answers to every scientific question, from the theory of the lever to the orbits of the heavenly bodies. But Greek civilization declined and the Romans conquered the Western world. One day just before Easter in A.D. 415, a woman was pulled from a chariot and killed by an ignorant mob. This scholar, devoted to geometry, to Pythagoras, and to rational thought, was the last famous scholar to work in the library at Alexandria before the descent of civilization into the thousand years of the Dark Ages.

Soon after civilization reemerged, so did geometry, but it was a new kind of geometry. It came from a man most civilized—he liked to gamble, sleep until the afternoon, and criticize the Greeks because he considered their method of geometric proof too taxing. To save mental labor, René Descartes married geometry and number. With his idea of coordinates, place and shape could be manipulated as never before, and number could be visualized geometrically. These techniques enabled calculus and the development of modern technology. Thanks to Descartes, geometric concepts such as coordinates and graphs, sines and cosines, vectors and tensors, angles and curvature, appear in every context of physics from solid state electronics to the large-scale structure of space-time, from the technology of transistors and computers to lasers and space travel. But Descartes's work also enabled a more abstract—and revolutionary—idea, the idea of curved space. Do all triangles really have angle sums of 180 degrees, or is that only true if the triangle is on a flat piece of paper? It is not just a question of origami. The mathematics of curved space caused a revolution in the logical foundations, not only of geometry but of all of mathematics. And it made possible Einstein's theory of relativity. Einstein's geometric theory of space and that extra dimension, time, and of the relation of space-time to matter and energy, represented a paradigm change of a magnitude not seen in physics since Newton. It sure *seemed* radical. But that was nothing, compared to the latest revolution.

One day in June 1984, a scientist announced that he had made a breakthrough in the theory that would explain everything from why subatomic particles exist, and how they interact, to the large-scale structure of space-time and the nature of black holes. This man believed that the key to understanding the unity and order of the universe lies in geometry—geometry of a new and rather bizarre nature. He was carried off the stage by a group of men in white uniforms.

It turned out the event was staged. But the sentiment and genius were real. John Schwarz had been working for a decade and a half on a theory, called string theory, that most physicists reacted to in much the same way they would react to a stranger with a crazed expression asking for money on the street. Today, most physicists believe that string theory is correct: the geometry of space is responsible for the physical laws governing that which exists within it.

The manifesto of the seminal revolution in geometry was written by a mystery man named Euclid. If you don't recall much of that deadly subject called Euclidean Geometry, it is probably because you slept through it. To gaze upon geometry the way it is usually presented is a good way to turn a young mind to stone. But Euclidean geometry is actually an exciting subject, and Euclid's work a work of beauty whose impact rivaled that of the Bible, whose ideas were as radical as those of Marx and Engels. For with his book, *Elements,* Euclid opened a window through which the nature of our universe has been revealed. And as his geometry has passed through four more revolutions, scientists and mathematicians have shattered theologians' beliefs, destroyed philosophers' treasured worldviews, and forced us to reexamine and reimagine our place in the cosmos. These revolutions, and the prophets and stories behind them, are the subject of this book.

THE STORY

OF EUCLID

What can you say about
space? How geometry
began describing the
universe and ushered in
modern civilization.

1. The First Revolution

UCLID was a man who possibly did not discover even one significant law of geometry. Yet he is the most famous geometer ever known and for good reason: for millennia it has been his window that people first look through when they view geometry. Here and now, he is our poster boy for the first great revolution in the concept of space—the birth of abstraction, and the idea of proof.

The concept of space began, naturally enough, as a concept of place, our place, earth. It began with a development the Egyptians and Babylonians called "earth measurement." The Greek word for that is *geometry,* but the subjects are not at all alike. The Greeks were the first to realize that nature could be understood employing mathematics—that geometry could be applied to reveal, not merely to describe. Evolving geometry from simple descriptions of stone and sand, the Greeks extracted the ideals of point, line, and plane. Stripping away the window-dressing of matter, they uncovered a structure possessing a beauty civilization had never before seen. At the climax of this struggle to invent mathematics stands Euclid. The story of Euclid is a story of revolution. It is the story of the axiom, the theorem, the proof, the story of the birth of reason itself.

2. The Geometry of Taxation

HE ROOTS of the Greek achievements sprouted in the ancient civilizations of Babylon and Egypt. Yeats wrote of Babylonian indifference, a trait that in mathematics, held them back from achieving greatness. Pre-Greek humanity noticed many clever formulae, tricks of calculation and engineering, but like our political leaders, they sometimes accomplished amazing feats with astonishingly little comprehension of what they were doing. Nor did they care. They were builders, working in the dark, groping, feeling their way, erecting a structure here, laying down stepping stones there, achieving purpose without ever achieving understanding.

They weren't the first. Human beings have been counting and calculating, taxing, and shortchanging each other since well before recorded times. Some alleged counting tools dating back to 30,000 B.C. might just be sticks decorated by artists with intuitive mathematical sensibilities. But others are intriguingly different. On the shores of Lake Edward, now in the Democratic Republic of Congo, archeologists unearthed a small bone, 8,000 years old, with a tiny piece of quartz stuck in a groove at one end. Its creator, an artist or mathematician—we'll never know for sure—cut three columns of notches into the bone's side. Scientists believe this bone, called the Ishango bone, is probably the earliest example ever found of a numerical recording device.

The thought of performing operations on numbers was much slower in coming because performing arithmetic requires a certain degree of abstraction. Anthropologists tell us that among many tribes, if two hunters fired two arrows to fell two gazelles, then got two hernias lugging them back toward camp, the word used for "two" might be different in each case. In these civilizations, you really couldn't add apples and oranges. It seems to have taken many thousands of years for humans to discover that these were all instances of the same concept: the abstract number, 2.

The first major steps in this direction were taken in the sixth millennium B.C., when the people of the Nile Valley began to turn away from nomadic life and focus on cultivating the valley. The deserts of northern Africa are among the driest and most barren spots in the world. Only the Nile River, swollen with equatorial rains and melted snow from the Abyssinian highlands, could, like a god, bring life and sustenance to the desert. In ancient times, in mid-June each year, the Nile Valley, dry and desolate and dusty, would feel the river drive forward and rise, filling up its bed, spreading fertile mud over the countryside. Long before the classical Greek writer Herodotus described Egypt as "the gift of the Nile," Ramses III left an account indicating how the Egyptians worshipped this god, the Nile, called *Hapi*, with offerings of honey, wine, gold, turquoise—all that the Egyptians valued. Even the name, "Egypt," means "black earth" in the Coptic language.

● ■ ▲

Each year, the inundation of the valley lasted four months. By October, the river would begin to shrivel and shrink until the land had baked dry once more by the following summer. The eight dry months were divided into two seasons, the *perit* for cultivation and the *shemu* for harvesting. The Egyptians began to establish settled communities built on mounds that, during the floods, became tiny islands joined by causeways. They built a system of irrigation and grain storage. Agricultural life became the basis for the Egyptian calendar and Egyptian life. Bread and beer became their staples. By 3500 B.C., the Egyptians had mastered minor industry, such as crafts and metalworking. Around that time, they also developed writing.

The Egyptians had always had death, but with wealth and settlement, they now also had taxes. Taxes were perhaps the first imperative for the development of geometry, for although in theory the Pharaoh owned all land and possessions, in reality temples and even private individuals owned real estate. The government assessed land taxes based on the height of the year's flood and the surface area of the holdings. Those who refused to pay might be beaten into submission on the spot by the police. Borrowing was possible but the interest rate was based on a "keep it simple" philosophy: 100 percent per year. Since much was at stake, the Egyptians developed fairly reliable, if tortuous, methods of calculating the area of a square, rectangle, and trapezoid. To find the area of a circle, they approximated it by a square with sides equal to eight-ninths the diameter. This is equivalent to using a value of 256/81, or 3.16, for pi, an overestimate, but off by only 0.6 percent. History does not record whether taxpayers griped about the inequity.

The Egyptians employed their mathematical knowledge to impressive ends. Picture a windswept, desolate desert, the date, 2580 B.C. The architect had laid out a papyrus with the plans for your structure. His job was easy—square base, triangular faces—and, oh yeah, it has to be 480 feet high and made of solid stone blocks weighing over 2 tons each. You were charged with overseeing completion of structure. Sorry, no laser sight, no fancy surveyor's instruments at your disposal, just some wood and rope.

As many homeowners know, marking the foundation of a building or the perimeter of even a simple patio using only a carpenter's square and measuring tape is a difficult task. In building this pyramid, just a degree off from true, and thousands of tons of rocks, thousands of person-years later, hundreds of feet in the air, the triangular faces of your pyramid miss, forming not an apex but a sloppy four-pointed spike. The Pharaohs, worshipped as gods, with armies who cut the phalluses off enemy dead just to help them keep count, were not the kind of all-powerful deities you would want to present with a crooked pyramid. Applied Egyptian geometry became a well-developed subject.

To perform their surveying, the Egyptians utilized a person called a *harpedonopta,* literally, a "rope stretcher." The harpedonopta employed three slaves, who handled the rope for him. The rope had knots in it at prescribed distances so that by stretching it taut with the knots as vertices, you could form a triangle with sides of given lengths—and hence angles of given measures. For instance, if you stretch a rope with knots at 30 yards, 40 yards, and 50 yards, you get a right angle between the sides of 30 and 40 degrees. (The word *hypotenuse* in Greek originally meant "stretched against"). The method was ingenious—and more sophisticated than it might seem. Today we would say that the rope stretchers formed not lines, but geodesic curves along the surface of the earth. We shall see that this is precisely the method, although in an imaginary, extremely small (technically, "infinitesimal") form, that we employ today to analyze the local properties of space in the field of mathematics known as differential geometry. And it is the Pythagorean theorem whose verity is the test of flat space.

While the Egyptians were settling the Nile, in the region between the Persian Gulf and Palestine another urbanization occurred. It began in Mesopotamia, the region between the Tigris and Euphrates Rivers, during the fourth millennium B.C. Sometime between 2000 and 1700 B.C. the non-Semitic people living just north of the Persian Gulf conquered their southern neighbors. Their victorious ruler, Hammurabi, named the combined kingdom after the city of Babylon. To the Babylonians we credit a system of mathematics considerably more sophisticated than that of the Egyptians.

Aliens gazing at earth through some super-telescope from 23,400,000,000,000,000 miles away can now observe Babylonian and Egyptian life and habits. For those of us stuck here, it is a bit harder to piece things together. We know Egyptian mathematics principally from two sources: the *Rhind Papyrus,* named for A. H. Rhind, who donated it to the British Museum, and the *Moscow Papyrus,* which resides in the Museum of Fine Arts in Moscow. Our best knowledge of the Babylonians comes from the ruins at Nineveh, where some 1,500 tablets were found. Unfortunately, none contained mathematical text. Luckily, a few hundred clay tablets were excavated in the region of Assyria, mostly from the ruins of Nippur and Kis. If combing through ruins is like searching a bookstore, these were the shops that had a math section. The ruins contained reference tables, textbooks, and other items that reveal much about Babylonian mathematical thought.

We know, for instance, that the Babylonian equivalent of an engineer would not just throw manpower at a project. To dig, say, a canal, he would note that the cross-section was trapezoidal, calculate the volume of dirt that had to be moved, take into account how much digging a man could do in a day, and come up with the number of man-days needed for the job. Babylonian moneylenders even calculated compound interest.

The Babylonians did not write equations. All their calculations were expressed as word problems. For instance, one tablet contained the spellbinder, "four is the length and five is the diagonal. What is the breadth? Its size is not known. Four times four is sixteen. Five times five is twenty-five. You take sixteen from twenty-five and there remains nine. What times what shall I take in order to get nine? Three times three is nine. Three is the breadth." Today, we would write "$x^2 = 5^2 - 4^2$." The disadvantage of the rhetorical statement of problems isn't as much the obvious one—its lack of compactness—but that the prose cannot be manipulated as an equation can, and rules of algebra, for instance, are not easily applied. It took thousands of years before this particular shortcoming was remedied: the oldest known use of the plus sign for addition occurs in a German manuscript written in 1481.

The excerpt above indicates that the Babylonians appear to have known the Pythagorean theorem, that for a right triangle the square of the hypotenuse is equal to the sum of the squares of the bases. As the rope stretcher's trick indicates, the Egyptians seem to have known this relation as well, but the Babylonian scribes filled their clay tablets with impressive tables of triplets of numbers exhibiting this dependence. They recorded low-lying triplets such as 3,4,5 or 5,12,13, but also others as large as 3456,3367,4825. The chances of finding a triplet that works by randomly checking threesomes of numbers is slim. For instance, in the first dozen numbers, 1, 2, . . . , 12, there are hundreds of ways to choose distinct triplets; of all these only the triplet 3,4,5 satisfies the theorem. Unless the Babylonians employed armies of calculators who spent their entire careers doing such calculations, we can conclude that they knew at least enough elementary number theory to generate these triplets.

Despite the Egyptians' accomplishments and the Babylonians' cleverness, their contributions to mathematics were limited to providing the later Greeks with a collection of concrete mathematical facts and rules of thumb. They were like classical field biologists patiently cataloguing species, not modern geneticists seeking to gain an understanding of how the organism develops and functions. For instance, though both civilizations knew the Pythagorean theorem, neither analyzed the general law that today we would write as $a^2 + b^2 = c^2$

(where c is the length of the hypotenuse of a right triangle, and a and b the lengths of the other two sides). They seem never to have questioned why such a relationship might exist, or how they might apply it to gain further knowledge. Is it exact, or does it only hold approximately? As a matter of principle, this is a critical question. In purely practical terms, who cares? Before the ancient Greeks came along, no one did.

Consider a problem that became the biggest headache in geometry in ancient Greece, but didn't bother the Egyptians or Babylonians at all. It's wonderfully simple. Given a square with sides measuring one unit in length, what is the length of the diagonal? The Babylonians calculated this as (converted to decimal notation) 1.4142129. That answer is accurate to three sexigesimal places (the Babylonians used a base sixty system). The Pythagorean Greeks realized the number cannot be written as a whole number or fraction, a situation that we today recognize as meaning that it is given by an endless string of decimals with no pattern: 1.414213562 . . . To the Greeks, this caused great trauma, a crisis of religious proportion, for the sake of which at least one scholar was murdered. Murdered for squealing about the value of the square root of 2? Why? The answer lies at the heart of Greek greatness.

3. Among the Seven Sages

HE DISCOVERY that mathematics is more than algorithms for calculating volumes of dirt or the magnitude of taxes is credited to a lone Greek merchant-turned-philosopher named Thales a bit more than 2,500 years ago. It is he who sets the stage for the great discoveries of the Pythagoreans, and eventually the *Elements* of Euclid. He lived at a time when, across the world, alarm clocks went off, in one way or another, waking the human mind. In India, Siddhartha Gautama Buddha, born around 560 B.C., began the spread of Buddhism. In China, Laotzu and his younger contemporary Confucius, born in 551 B.C., made intellectual progress of enormous consequence. In Greece, too, a Golden Age was beginning.

Near the west coast of Asia Minor, a river named Meander, the river from which the word *meander* is coined, spills into a dismal swampy plain in the country that today is Turkey. In the midst of that swamp, some 2,500 years ago, stood the most prosperous Greek city of its time, Miletus. It was then a coastal city, on a gulf now filled in by silt, in a region known as Ionia. Miletus was shut in by water and mountains, with only one convenient route to the interior, but at least four harbors, a center of maritime trade for the eastern Aegean. From here, vessels snaked their way south among the islands and peninsulas toward Cyprus, Phoenicia, and Egypt, or headed west to European Greece.

In this city, in the seventh century B.C., began a revolution in human thought, a mutiny against superstition and sloppy thinking that was to continue its development for nearly a millennium, and leave behind the foundations of modern reasoning.

Our knowledge of these groundbreaking thinkers is uncertain, often based on the biased writings of later scholars such as Aristotle and Plato, sometimes on contradictory accounts. Most of these legendary figures had Greek names, but they did not accept Greek myth. They were often persecuted, driven into exile, even suicide—at least according to the stories passed down about them.

Despite the differing accounts, it is generally agreed that in Miletus, around 640 B.C., a proud mother and father parented a baby boy they named Thales. Thales of Miletus has the honor of most often being named the world's first scientist or mathematician. Attaching this early date to these professions apparently does not threaten the primacy of that oldest profession, the sex business, as sections of padded leather designed for female sexual gratification were one of the items for which Miletus was known. We don't know whether Thales traded in those, or in salted fish, wool, or the other commodities for which Miletus was famous; but he was a wealthy merchant, and he used his cash to do what he pleased, retiring to devote himself to study and travel.

Ancient Greece comprised a number of small, politically independent political units, the city-states. Some were democratic, others controlled by a small aristocracy or a tyrannical king. Of Greek daily life, we know the most about Athens, but a citizen's life had many similarities throughout the Hellenes, and changed little over the few centuries following Thales, except during times of famine or war. The Greeks seemed to like socializing, at the barbershop, the temple, the marketplace. Socrates was a fan of the shoemaker's shop. Diogenes Laertius wrote of a cobbler, named Simon, who first introduced Socratic dialogues as a form of conversation. In the remains of a fifth-century B.C. shop, archeologists have unearthed a chip of a wine cup bearing the name "Simon."

The ancient Greeks also enjoyed dinner parties. In Athens, dinner would be followed by the symposium—literally, "together drinking." Revelers gulped diluted wine, discussing philosophy, singing songs, reciting jokes and riddles. Those failing at riddles, or committing various gaffes, were assessed punishments such as having to dance naked around the room. But if Greek partying is reminiscent of college life, so is their focus on knowledge. The Greeks valued inquiry.

Thales seems to have had the insatiable thirst for learning that characterized so many Greeks who shaped its Golden Age. In his travels to Babylon, he studied the science and mathematics of astronomy, and gained local fame by bringing this knowledge to Greece. One of Thales' legendary accomplishments was to predict the solar eclipse of 585 B.C. Herodotus tells us that it occurred during a battle, stopped the fighting, and brought on a lasting peace.

Thales also spent extended amounts of time in Egypt. The Egyptians had the expertise to build the pyramids, yet lacked the insight needed to measure their height. Thales sought theoretical explanations for the facts discovered empirically by the Egyptians. With such understanding, Thales could *derive* geometric techniques, one from another, or he could steal the solution for one problem from that of another because he had extracted the abstract principle from the particular practical application. He stunned the Egyptians by showing them how they could measure the height of the pyramids employing a knowledge of the properties of similar triangles. Thales later used a similar technique to measure the distance of a ship at sea. He became a celebrity in ancient Egypt.

In Greece, Thales was named by his contemporaries as one of the Seven Sages, the seven wisest men in the world. His feats were all the more impressive considering the primitive sense of mathematics possessed by the average person alive at that time. For instance, even centuries later, the great Greek thinker Epicurus still maintained that the sun was no huge ball of fire, but rather, "just as big as we see it."

Thales made the first steps toward the systemization of geometry. He was the first to prove geometric theorems of the kind Euclid would gather in his *Elements* centuries later. Realizing that rules were needed for determining what might validly follow from what, Thales also invented the first system of logical reasoning. He was the first to consider the concept of congruence of spatial figures, that two figures in a plane can be considered equal if you can slide and rotate one to coincide exactly with the other. Extending the idea of equality from number to spatial objects was a giant leap in the mathematization of space. It is also not as obvious as it may seem to those of us indoctrinated to this early in our school days. In fact, as we will see, it involves the assumption of homogeneity, that a figure neither warps nor alters size as it moves, which is not true in all spaces, including our own physical space. Thales kept the Egyptian name "earth measurement" for his mathematics, but being Greek, used the Greek word *geometry.*

Thales asserted that via observation and reasoning we should be able to explain all that happens in nature. He eventually came to the revolutionary conclusion that nature follows regular laws. Thunderclaps are not the loud noises made by angry Zeus. There has to be a better explanation, obtained by observation and reasoning. And in mathematics, conclusions about the world should be verified via rules, not guesses and observation.

Thales also addressed the concept of physical space. He recognized that all matter in the world, despite its vast variety, must be intrinsically the same stuff. In the absence of any evidence, it was an amazing leap of intuition. The next natural question was, of course, what is this fundamental stuff? Here, living in a city of harbors, intuition led Thales to choose water. Ironically, Thales' student and fellow Milesian, Anaximander, came by a comparable leap of intuition to the idea of evolution, and for the lower animal from which humans evolved, chose the fish.

When Thales was a frail old man, fearful of his own senility, he met Euclid's most important forerunner, Pythagoras of Samos. Samos was a city on a large island of the same name, in the Aegean Sea, not far from Miletus. A visitor to the island today can still find some shattered columns, and basalt remains of a theater overlooking the site of its ancient harbor. In Pythagoras' day, it flourished. When Pythagoras was eighteen, his father died. His uncle gave him some silver and a letter of introduction, and sent him off to visit the philosopher Pherecydes, on the nearby island of Lesbos, the island from which the term *lesbian* is derived.

According to legend, Pherecydes had studied the secret books of the Phoenicians, and introduced to Greece the belief in immortality of the soul and reincarnation, which Pythagoras embraced as cornerstones of his religious philosophy. Pythagoras and Pherecydes became lifelong friends, but Pythagoras did not stay long on Lesbos. By the time he was twenty, Pythagoras journeyed to Miletus, where he met Thales.

The historical picture is one of a young boy with long stringy hair, dressed not in the traditional Greek tunic, but instead clad in pants, a kind of ancient hippy, visiting the famous old sage. Thales by then was a man cognizant that his earlier brilliance had dimmed considerably. Seeing perhaps a glimmer of his own youth in the boy, he apologized for his diminished mental state.

We know little of what Thales actually said to Pythagoras, but we do know he was a great influence on the young genius. Years after Thales' death, Pythagoras would sometimes be found sitting at home, singing songs of praise to the departed visionary. All ancient accounts of the meeting agree on one thing: Thales gave Pythagoras the Horace Greeley treatment, but instead of telling him to go west, young man, Thales recommended Egypt.

4. The Secret Society

YTHAGORAS took Thales up on his recommendation to go to Egypt, but there, Pythagoras found no poetry in Egyptian mathematics. Geometric objects were physical entities. A line was the rope the harpedonopta tugged, or the edge of a field. A rectangle was the boundary of a plot of land, or the face of a stone block. Space was mud, soil, and air. To the Greeks, not the Egyptians, goes the credit for the idea that brings romance and metaphor to mathematics: that space can be a mathematical abstraction, and, just as important, that the abstraction can apply to many different circumstances. Sometimes a line is just a line. But the same line can represent the edge of a pyramid, the boundary of a field, or the path the crow flies. Knowledge about one transfers to the other.

According to legend, Pythagoras was walking by a blacksmith's shop one day, when he heard the tone of various hammers pounding on a heavy anvil. This made him think. After some experimentation with strings, he discovered harmonic progressions, and the relationship between the length of a vibrating string and the pitch of the musical note it produces. A string twice as long, for instance, produces a note with half the pitch. A simple observation, but a deep and revolutionary act, it is often considered the first example in history of an empirical discovery of a natural law.

Millions of years ago, somebody eeked or hrmphed and another somebody uttered immortal words, now lost, but what must have meant something like "I know what you mean." The idea of language had arrived. In science, Pythagoras' law of harmonics represents an equal milestone, the first example of the physical world phrased in mathematical terms. In his day, it must be remembered, the mathematics of simple numerical phenomena was unknown. For instance, to the Pythagoreans it was a revelation that multiplying the dimensions of a rectangle gave you its area.

For Pythagoras, much of the intrigue of mathematics came from the many numerical patterns he and his followers discovered. The Pythagoreans envisioned the integers as pebbles or dots, which they laid out in certain geometric patterns. They found that some numbers can be formed by laying the pebbles equally spaced in two columns of two, three of three, and so on, so that the array forms a square. The Pythagoreans called any number of pebbles you can arrange this way "square numbers," which is why we call these numbers "squares" today: 4, 9, 16, etc. Other numbers, they found, could be formed by laying out the pebbles in columns of one, two, three and so on, to form triangles: 3, 6, 10, etc.

The properties of square and triangular numbers fascinated Pythagoras. For instance, the second square number, 4, is equal to the sum of the first two odd numbers, 1 + 3. The third, 9, is equal to the sum of the first three odd numbers, 1 + 3 + 5, and so on. (This is also true for the first square, 1 = 1.) While the square numbers all equal the sum of consecutive odd numbers, Pythagoras noticed that in the same way the triangular numbers are sums of all consecutive numbers, both even and odd. And square and triangular numbers are related: if you add a triangular number to the preceding or to the next triangular number, you get a square number.

The Pythagorean theorem, too, must have seemed magical. Imagine ancient scholars scrutinizing triangles of every ilk, not just the rare right triangle, measuring their angles and sides, rotating and comparing them. If such an investigation occurred today, universities might well have a discipline devoted to it. "My son is on the math faculty at Berkeley," some

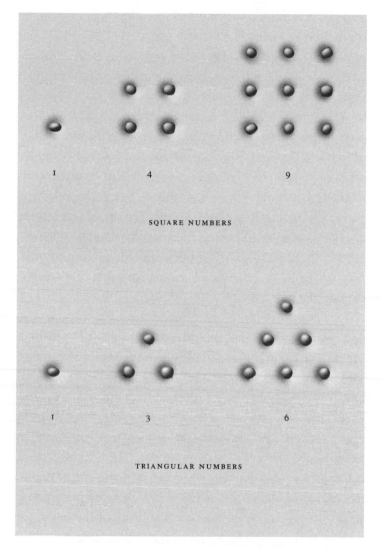

1   4   9

SQUARE NUMBERS

1   3   6

TRIANGULAR NUMBERS

PHYTHAGORAS' PEBBLE PATTERNS

proud mother would say. "He's a professor of triangles." One day her boy notices a peculiar regularity, that in every right triangle the square of the length of the hypotenuse equals the sum of the squares of the other two sides. It proves true for big ones, small ones, fat ones, short ones, for every right triangle ever measured, yet not for any other type of triangle. It's a discovery that would surely rate a headline on the front page of the *New York Times:* "Surprising Regularity Discovered in the Right Triangle," and in smaller print, *"Applications Still Years Away."*

Why should the sides of all right triangles always obey such a simple relationship? The Pythagorean theorem can be proved using a kind of geometric multiplication Pythagoras often employed. We don't know if this is how he proved this theorem, but proving it this way is revealing because it is purely geometric. Today, simpler proofs exist, which rely on algebra or even trigonometry, neither of which were developed in Pythagoras' day. But the geometric proof isn't difficult; it's really just a twisted mathematician's version of a connect-the-dots activity.

To prove the Pythagorean theorem the geometric way, the only computational fact you'll need is that the area of a square is equal to the square of the length of one of its sides. This is just a modern restatement of Pythagoras' pebble analogy. Given any right triangle, the goal is to form three squares from it: one square whose sides each are equal in length to the hypotenuse; and two other squares whose sides correspond in length to the triangle's other two sides. The area of each of these three squares is then the square of the length of one of the triangle's sides. If we can show that the hypotenuse square's area is equal to the combined area of the other two squares, then we will have proved the Pythagorean theorem.

To make things simple, let's give the sides of the triangle names. The hypotenuse already has a name, albeit a lengthy one, so we'll keep that, except we will capitalize it to distinguish the name of our particular line, Hypotenuse, from the term *the hypotenuse.* Let's call the other two sides of the triangle Alexei and Nicolai. Coincidentally, these are the names of the author's two sons. At the time of this writing, Alexei is the longer, and Nicolai is the shorter, so let's use that convention in naming the sides of the triangle (the proof works equally well with sides of equal length). We begin the construction by drawing a square whose sides are each the combined length of Alexei and Nicolai. Next, draw a dot on each side, dividing each side into one segment with Alexei's length, and another with Nicolai's length, and connect the dots. There are different ways to do this. The two ways we are interested in are illustrated in the figure on page 22. One results in a square whose sides match Hypotenuse, plus four "leftover" triangles. The other results in two squares whose sides match Alexei and Nicolai, plus two leftover rectangles which may be cut along their diagonals to form four leftover triangles identical to the leftovers we got doing it the other way.

The rest is just accounting. The two subdivided squares have identical areas, so after discarding the four leftover triangles from each, the real estate that remains in one square remains equal to that in the other. But in one figure that area is the square of the length of Hypotenuse, and in the other it is the sum of the squares of Alexei's and Nicolai's length. So we have proved the theorem!

Impressed by such new triumphs of knowledge, one of Pythagoras' disciples wrote that "were it not for number and its nature, nothing that exists would be clear to anybody." A reflection of their fundamental philosophy, the Pythagoreans invented the term *mathematics,* from the Greek word *mathema,* which meant "science." The word's origin reflects the close connection between the two subjects, though today there is a sharp distinction between mathematics and science, a distinction, as we shall see, that didn't become clear until the nineteenth century.

There is also a distinction between intelligent talk and blather, a distinction that Pythagoras did not always make. Pythagoras' awe of numerical relations swept him into form-

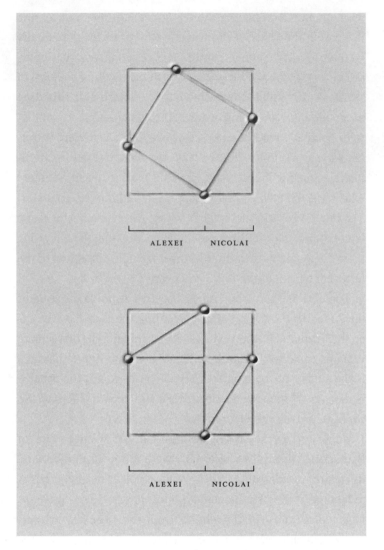

PYTHAGORAS' THEOREM

ing many mystic numerological beliefs. He was the first to divide numbers into the categories "odd" and "even," but he took the extra step of personifying them: the odd he called "masculine," the even, "feminine." He associated specific numbers with ideas, such as the number 1 with reason, 2 with opinion, 4 with justice. Since 4 in his system was represented by a square, the square was associated with justice, the origin of the expression we still use today, "a square deal." In the interests of giving Pythagoras a square deal, one must recognize that it is easier to judge the brilliant from the blather with the perspective of a couple thousand years.

Pythagoras was a charismatic figure and a genius, but he was also a good self-promoter. In Egypt, he not only learned Egyptian geometry but became the first Greek to learn Egyptian hieroglyphics, and eventually became an Egyptian priest, or the equivalent, initiated into their sacred rites. This gave him access to all their mysteries, even to the secret rooms in their temples. He remained in Egypt for at least thirteen years. When he left, it wasn't of his own volition—the Persians invaded and took him prisoner. Pythagoras landed in Babylon, where he eventually obtained his freedom, and gained a thorough knowledge of Babylonian mathematics as well. He finally returned to Samos, at the age of fifty. By the time Pythagoras made it back to his homeland, he had synthesized the philosophy of space and mathematics he was intent on preaching; all he needed were some followers.

His knowledge of hieroglyphics led many Greeks to believe he had special powers. He encouraged tales that set him apart from normal citizens. One of the more bizarre stories had him attacking a poisonous snake and biting the snake to death. Another describes a thief who broke into Pythagoras' home and saw such bizarre things that he fled empty-handed, refusing ever to reveal the strange things he saw. Pythagoras had a golden birthmark on his thigh, which he displayed as a sign of divinity. The people of Samos did not prove extremely susceptible to his preachings, so Pythagoras soon left for a less sophisticated home, Croton, an Italian city colonized by Greeks. There, he established his "society" of followers.

The life and legend that developed around Pythagoras in

many ways parallels that of a later charismatic leader, Jesus Christ. It is hard to believe that the myths told about Pythagoras did not influence the creation of some of the later stories about Christ. Pythagoras, for instance, was believed by many to be the son of God, in this case, Apollo. His mother was called Parthenis, which means "virgin." Before traveling to Egypt, Pythagoras lived the life of a hermit on Mount Carmel, like Christ's solitary vigil on the mountain. A Jewish sect, the Essenes, appropriated this myth and is said to have later had a connection to John the Baptist. There is also a myth that Pythagoras returned from the dead, although, according to the story, Pythagoras faked this by hiding in a secret underground chamber. Many of Christ's miraculous powers and deeds were first ascribed to Pythagoras: he is said to have appeared in two places at once; he could calm waters and control winds; he was once greeted by a divine voice; he was believed to have the ability to walk on water.

Pythagoras' philosophy also had some similarities to that of Christ. For instance, he preached that you should love your enemies. But in philosophy, he was closer to his contemporary Siddhartha Gautama Buddha (c. 560–480 B.C.). Both believed in reincarnation, possibly as an animal, so even an animal could be inhabited by what was once a human soul. Thus, both placed a high value on all life, opposing the common practice of animal sacrifice and preaching strict vegetarianism. According to one story, Pythagoras once stopped a man from beating a dog by telling the man he recognized the canine as an old friend of his, reincarnated.

Pythagoras felt that possessions got in the way of the pursuit of divine truths. Greeks of that period would sometimes wear wool, and often used colors on their garments. Well-to-do men occasionally tossed a capelike mantle over their shoulders, fastened with a gold pin or brooch, proudly displaying their wealth. Pythagoras rejected luxury and banned his followers from any clothing except that made from simple white linen. They earned no money, but relied on the charity of the Croton populace and perhaps the wealth of some of his followers, who pooled their possessions and lived in a communal lifestyle. It is hard to determine the nature of his organization because, in their attitudes and customs, people of that time and place were so different. For instance, two of the ways Pythagoras's set distinguished themselves from the ordinary were by not urinating in public and not having sex in front of others.

Secrecy played an important role in Pythagorean society, perhaps based on his experience with the secret practices of the Egyptian priesthood. Or perhaps, the motivation was a desire to avoid the trouble that would be caused by revealing revolutionary ideas that might stir opposition. One of Pythagoras' discoveries became such a secret that according to legend, the Pythagoreans forbade its revelation on penalty of death.

Recall the problem of determining the length of the diagonal of the unit square. The Babylonians calculated it to six decimal places, but for the Pythagoreans, this was not good enough. They wanted to know its exact value. How could you pretend to know anything about the space inside a square if you didn't know that? The trouble was, though they could achieve better and better approximations, none of the numbers they produced turned out to be the exact answer. But the Pythagoreans were not easily daunted. They had the imagination to ask themselves, does this number even exist? They concluded that it does not, and they had the ingenuity to prove it.

Today, we know that the length of the diagonal is equal to the square root of 2, an irrational number. That means that it cannot be written in decimal form with a finite number of digits, or equivalently, that it cannot be represented as a whole

number or fraction, the only kind of numbers the Pythagoreans knew. Their proof that the number does not exist was actually a proof that it cannot be written in fractional form.

Clearly, Pythagoras had a problem. The fact that the length
of the diagonal of a square could not be expressed as any
number was not good for a visionary who preaches that number is everything. Should he alter his philosophy: number is
everything, except for the certain geometric magnitudes
which we find really mysterious?

Pythagoras could have pushed up the invention of the real
number system by many centuries, had he done a simple
thing: given the diagonal a name, say, d, or even better, $\sqrt{2}$,
and considered it some new kind of number. Had he done
that, he might have pre-empted Descartes's coordinate revolution, for, absent a numerical representation, the need to describe this new type of number begged for the invention of
the number line. Instead, Pythagoras retreated from his
promising practice of associating geometric figures with
numbers, and proclaimed that some lengths cannot be expressed as a number. The Pythagoreans called such lengths
alogon, "not a ratio," which we today translate as "irrational."
The word *alogon* had a double meaning, though: it also meant
"not to be spoken." Pythagoras had solved his dilemma with a
doctrine that would have been hard to defend, so, in keeping
with his overall doctrine of secrecy, he banned his followers
from revealing the embarrassing paradox. Not all obeyed. According to legend, one of his followers, Hippasus, did reveal
the paradox. Today people are murdered for many reasons—
love, politics, money, religion—but not because somebody
squealed about the square root of 2. To the Pythagoreans,
though, mathematics was a religion, so when Hippasus broke
the oath of silence, he was assassinated.

Resistance to irrationals continued for thousands of years.
In the late nineteenth century, when the gifted German math-

ematician Georg Cantor did groundbreaking work to put them on firmer footing, his former teacher, a crab named Leopold Kronecker who "opposed" the irrationals, violently disagreed with Cantor and sabotaged his career at every turn. Cantor, unable to tolerate this, had a breakdown and spent his last days in a mental institution.

Pythagoras also ended his life in trouble. Around 510 B.C., some Pythagoreans traveled to a nearby city named Sybaris, apparently seeking followers. Few details of their mission survive, except that they were murdered. Later, a faction of Sybarites fled to Croton, escaping from a tyrant, Telys, who had recently gained power in the city. Telys demanded their return. Pythagoras broke one of his cardinal rules: Stay out of politics. He persuaded the Crotonites not to deport the exiles. A war ensued, which Croton won, but to Pythagoras, the damage was done. He now had political enemies. Around 500 B.C., they attacked his group. Pythagoras fled. It is not clear what happened to him after that: most sources say he committed suicide; others say he lived out his years quietly and died around the age of one hundred.

The Pythagorean society continued for some time after the attack, until another attack, around 460 B.C., slaughtered all but a couple of his followers. His teachings survived in some form until about 300 B.C. They were revived by the Romans, in the first century before Christ, and became a dominating force within the budding Roman Empire. Pythagoreanism became an influence in many religions of that time, such as Alexandrian Judaism, the aging ancient Egyptian religion, and, as we have seen, in Christianity. In the second century A.D., Pythagorean mathematics, in association with the School of Plato, received new impetus. Pythagoras' intellectual descendants were again squelched by Justinian, the eastern Roman emperor, in the fourth century A.D. The Romans hated the long hair and beards of Pythagoras' Greek philosopher descendants, and their use of drugs, such as opium, not to mention their un-Christian beliefs. Justinian closed the academy and forbade the teaching of philosophy. Pythagoreanism flickered for a couple more centuries, then disappeared into the Dark Ages around A.D. 600.

5. Euclid's Manifesto

ROUND 300 B.C., on the southern shore of the Mediterranean Sea, a little west of the Nile in Alexandria, lived a man whose work has had influence rivalling that of the Bible. His approach informed philosophy, and defined the nature of mathematics until well into the nineteenth century. His work was an integral part of higher education for most of that time, and continues so today. The recovery of his work was a key to the renewal of European civilization in the Middle Ages. Spinoza emulated him. Abraham Lincoln studied him. Kant defended him.

The name of this man was Euclid. Of his life, virtually nothing is known. Did he eat olives? Did he see plays? Was he tall or short? History answers none of these questions. All we know is that he opened a school in Alexandria, had brilliant students, scorned materialism, seemed to be a pretty nice guy, and wrote at least two books. One of them, a lost book on conics, the study of curves generated by the intersection of a plane and a cone, formed the basis of later momentous work by Apollonius which substantially advanced the sciences of navigation and astronomy.

His other famous work, *Elements,* is one of the most widely read "books" of all time. The *Elements* has a history deserving of *The Maltese Falcon.* First, it is actually not a book, but a series of thirteen rolls of parchment. None of the originals survive, but instead were passed down through a series of later editions, and in the Dark Ages almost disappeared completely. The first four scrolls of Euclid's work are not the original *Elements* anyway: a scholar named Hippocrates (not the physician of the same name) wrote a work

29

called the *Elements* around 400 B.C., which is believed to have been the source for most of what appears in those. None of the contents of *Elements* is credited. Euclid made no claims of originality regarding any of the theorems. He saw his role as organizing and systematizing the Greek understanding of geometry. He was the architect of the first comprehensive account of the nature of two-dimensional space via pure thought, with no reference to the physical world.

The most important contribution of Euclid's *Elements* was its innovative logical method: first, make terms explicit by forming precise definitions and so ensure mutual understanding of all words and symbols. Next, make concepts explicit by stating explicit axioms or postulates (these terms are interchangeable) so that no unstated understandings or assumptions may be used. Finally, derive the logical consequences of the system employing only accepted rules of logic, applied to the axioms and to previously proved theorems.

Picky, picky, picky. Why be so insistent on proving every tiny assertion? Mathematics is a vertical edifice that, unlike a tall building, will topple if just one mathematical brick is corrupt. Allow even the most innocuous fallacy into the system and you can't trust anything. In fact, a theorem of logic states that if any false theorem is allowed into a logical system, no matter what it pertains to, then you will be able to use it to prove that 1 equals 2. According to legend, a skeptic once cornered logician Bertrand Russell, attempting to attack this sweeping theorem (though actually he was speaking of the converse). "Okay," barked the doubter, "if I allow that one equals two, then prove that you're the pope." Russell is said to have thought for the tiniest moment, and then replied, "The pope and I are two, therefore the pope and I are one."

Proving every assertion means in particular that intuition, though a valuable guide, must be checked at proof's door. The phrase "It is intuitively obvious" is not proper justification for a step in a proof. We are all far too fallible for that. Imagine rolling out a ball of yarn along the earth's equator, all 25,000 miles of it. Now imagine doing the same a foot above the equator. How much more yarn do you need—500 feet, 5,000 feet? Let's make it easier. Imagine unrolling two more balls, this time one on the surface of the sun, the other a foot above. To which ball must you add more yarn when you move out a foot, the earth's or the sun's? Intuition tells most of us it is the sun's, but the answer is, you've added exactly the same to each, 2 pi feet, or about 6 feet 3 inches.

Long ago there was a television show called *Let's Make a Deal*. A contestant would face three stages, concealed by curtains. One stage would contain an item of great value, like a car; the other two, booby prizes. Let's say the contestant chose curtain two. The host would then have one of the other curtains opened, say curtain three. Suppose curtain three revealed a booby prize, so the real prize is behind curtain one or your chosen curtain, two. The host would then ask you if you'd like to change your choice, in this case to choose number one instead of number two. Do you do it? It seems, intuitively, that your chances are the same, fifty-fifty, regardless. That would be the case if you had no other information, but you do; you have the history of your earlier choice and the host's actions. A careful analysis of all the possibilities from your initial choice onward, or the application of the appropriate formula, called Bayes' theorem, will reveal that your chances are better if you change your selection. There are many examples in mathematics where intuition fails and only deliberate formal reasoning will reveal the truth.

Exactness is another property required in mathematical proof. An observer might measure the diagonal of the unit square as 1.4, or refine her instrument and obtain 1.41 or 1.414, and though we might be tempted to accept such approximations as good enough, what such approximations

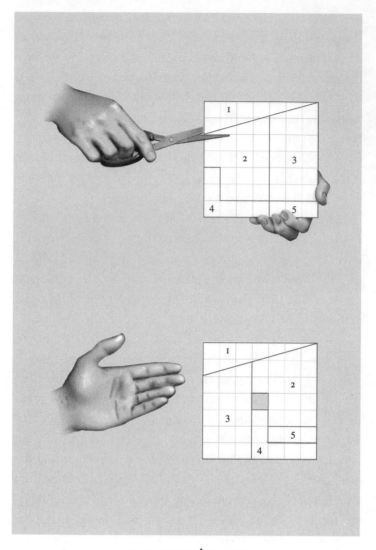

PAUL CURRY'S TRICK

could never reveal is the revolutionary insight that the length is irrational.

Tiny quantitative changes can have large qualitative consequences. Think about state lotteries. Hopeful losers often shrug, "You can't win if you don't play." That is certainly true. But it is equally true that, within a tiny fraction of a percent, your chances of winning are the same whether you buy a lottery ticket or not. What would happen if the lottery commission announced it had decided to round off your chances of winning from 0.00001 percent to zero? It would be a small change, but it would have a large consequence in their revenue stream.

A trick invented by Paul Curry (see opposite page), an amateur magician living in New York City, provides a good geometric example of this effect. Take a square piece of paper on which is drawn a seven-by-seven grid of smaller squares. Cut the large square into five pieces and then rearrange the pieces as shown in the figure. The result is a "square donut," a square of the same size as the original, with one of the small squares missing from its center. What happened to the missing area? Have we proved a theorem that the whole square and the donut have the same area?

The answer is that when the fragments are pieced back together, there is just a bit of overlap, so the figure is a bit of a cheat—or let's say, an approximation. The second from the top row of squares has just a bit of extra height, so the large square is $\frac{1}{49}$ taller than it should be—exactly enough to account for the area of the missing square. But if we were constrained to measuring lengths to a precision of 2 percent, we couldn't tell the difference between the two constructions and we might be tempted to conclude the magical result that the area of the square and the "donut square" are equal.

Do such small deviations play a role in actual theories of space? One of Albert Einstein's key guides in creating his general theory of relativity, his revolutionary theory of curved space, was a deviation from classical Newtonian theory in the perihelion of Mercury. According to Newton's theory, planets move in perfect ellipses. The point at which a planet is closest to the sun is called the *perihelion point,* and if Newton's theory is correct, a planet should return to precisely the same perihelion as it orbits the sun each year. In 1859, Urbain-Jean-Joseph Leverrier announced in Paris that he had found that the perihelion of Mercury actually migrates by an extremely small amount—certainly an amount of no practical consequence—38 seconds a century. Yet the deviation had to be due to something. Leverrier called it "a grave difficulty, worthy of attention by astronomers." In 1915, Einstein had developed his theory far enough to calculate Mercury's orbit, and he found agreement with the tiny deviation. According to one biographer, Abraham Pais, it was "the high point in his scientific life. He was so excited that for three days he could not work." Tiny as it is, the deviation had required nothing less than the fall of classical physics.

Euclid's aim was that his system be free of unrecognized assumptions based on intuition, of guesswork and of inexactness. He stated twenty-three definitions, five geometric postulates, and five additional postulates he called "common notions." From this foundation, he proved 465 theorems—essentially all the geometric knowledge of his day.

Euclid's definitions included terms like *point, line* (which in his definition could be curved), *straight line, circle, right angle, surface,* and *plane.* He defined some of these terms quite precisely. Parallel lines, he wrote, are "straight lines which, being in the same plane, and being produced indefinitely in both directions, do not meet one another in either direction."

A circle, he wrote, is "a plane figure contained by one line [i.e., curve] such that all straight lines falling upon it from one point amongst those lying within the circle—called the center—are equal to one another." For the right angle, Euclid wrote: "When a straight line set up upon a straight line makes the adjacent angles equal to each other, each of the equal angles is a right angle."

Some of Euclid's other definitions, such as those for point and line, are vague and almost useless: a straight line is "that which lies evenly with the points on itself." This definition may have come from the building trade, where you checked a line for straightness by closing an eye and peering along its length. To understand it, you must already have had the image of a line. A point is "that which has no part," another definition that borders on meaningless.

Euclid's common notions were more elegant. They were non-geometric assertions of logic he apparently thought to be common sense, as opposed to the postulates, which are specific to geometry. It was a distinction made previously by Aristotle. By explicitly exposing these intuitive assumptions, he was essentially adding to his postulates, yet he apparently felt the need to differentiate them from his purely geometric assertions. It is a testament to his depth of thought that he saw the need to make such statements at all:

1. Two things which are both equal to a third thing are also equal to each other.

2. If equals are added to equals, the wholes are equal.

3. If equals be subtracted from equals, the remainders are equal.

4. Things which coincide with one another are equal to one another.

5. The whole is greater than the part.

These preliminaries aside, the geometric content of the foundation of Euclid's geometry lies in his five postulates. The first four are simple and can be stated with a certain grace. In modern terms, they are:

1. Given any two points, a line segment can be drawn with those points as its endpoints.

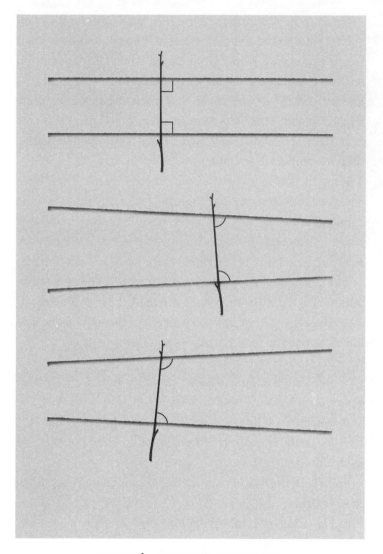

EUCLID'S PARALLEL POSTULATE

2. Any line segment can be extended indefinitely in either direction.

3. Given any point, a circle with any radius can be drawn with that point at its center.

4. All right angles are equal.

Postulates 1 and 2 seem to coincide with our experience. We feel we know how to draw a line segment from point to point, and we have never run into any barriers where space ends, preventing us from extending line segments. His third postulate is a bit more subtle—part of what it implies is that distance in space is defined in such a way that a line segment's length does not change when we move it from one place to another as we trace out a circle. His fourth postulate sounds simple and obvious. To understand the subtleties involved, recall the definition of the right angle: it is the angle made when one line intersects another in such a way that the angles it thus forms on both sides are equal. We have seen this many times: one line is perpendicular to the other, and the angles it forms on either side at the intersection both measure 90 degrees. But the definition alone doesn't assert this; it doesn't even stipulate that the measure of the angles is always the same number. We might imagine a world in which the angles might equal 90 degrees if the lines intersect at one given point, but if they intersect elsewhere, the angle equals some other number. The postulate that all right angles are equal guarantees that this cannot happen. It means, in a sense, that a line looks the same all along its length, a kind of straightness condition.

Euclid's fifth postulate, called the *parallel postulate*, does not sound as obvious or intuitive as the others. It is Euclid's own invention, not part of the great body of knowledge that he was chronicling. Yet he apparently did not like this postulate, as he appeared to avoid its use whenever possible. Later mathematicians did not like it either, feeling it was not simple enough for a postulate, and ought to be provable as a theorem. Here it is, in a form close to Euclid's original:

5. Given a line segment that crosses two lines in a way that the sum of inner angles on the same side is less than two right angles, then the two lines will eventually meet (on that side of the line segment).

The parallel postulate (p. 36) gives a test for deciding whether two coplanar lines are converging, parallel, or diverging. It helps to have a diagram to see this.

There are many different but equivalent formulations of the parallel postulate. One which makes what this postulate says about space especially clear is:

Given a line and an external point (a point not on the line), there is exactly one other line (in the same plane) that passes through the external point and is parallel to the given line.

The parallel postulate could be violated in two possible ways: there might be no such thing as parallel lines, or there might exist more than one parallel line through some external point.

● ■ ▲

Draw a line on a piece of paper, and a dot somewhere not on the line. Does it seem possible you cannot draw any parallel line through the dot? Does it seem possible to draw more than one? Does the parallel postulate describe our world? Could a geometry in which it is violated be mathematically consistent? These last two questions eventually led to a revolution in intellectual thought, the former in our view of the universe, the latter in our understanding of the nature and meaning of mathematics. But for 2,000 years, there was hardly another idea in any field of human knowledge that was more universally accepted than the "fact" expressed by Euclid's postulate, that only and exactly one parallel exists.

6. A Beautiful Woman, a Library, and the End of Civilization

UCLID was the first great mathematician of a long, and unfortunately doomed, line of scholars to work in Alexandria. The Macedonians, a Greek people living in the north of mainland Greece, began the conquest and unification of Hellenic lands under Philip II of Macedonia in 352 B.C. After a decisive defeat, Athenian leaders accepted peace on Philip's terms in 338 B.C., effectively ending the independence of Greek city-states. Only two years later, while attending a state ceremony in which his own statue was displayed as a new Olympian god, Philip II fell victim to a bad hire: he was stabbed to death by one of his own bodyguards. His son, Alexander, as in Alexander the Great, then twenty, took command.

Alexander placed a great value on knowledge, perhaps due to his liberal education in which geometry played an important role. He respected foreign cultures, though apparently not their independence. He soon conquered the rest of Greece, Egypt, and the Near East as far as India. He encouraged cross-cultural communication and intermarriage, marrying a Persian woman himself. Not content to lead by example, he ordered leading Macedonian citizens to also marry Persian women.

In 332 B.C., in the center of his empire, the cosmopolitan Alexander began constructing his lavish capital, Alexandria. In this respect, a kind of Walt Disney of ancient times, he envisioned a carefully laid out, "planned" metropolis. It was to be a center of culture, trade, and government. Even in the layout of its wide boulevards, he seemed to be making a mathematical statement: his architect laid them out in a grid pattern, a curious anticipation of the coordinate geometry not invented for another eighteen centuries.

Alexander died of an unknown illness nine years after construction began, before his grand city was finished. His empire disintegrated, but Alexandria was eventually completed. Its geometry was felicitous because the city became the center of Greek mathematics, science, and philosophy after a former Macedonian general named Ptolemy took over the Egyptian portion of Alexander's empire. Ptolemy's son, named, creatively, Ptolemy II, later took over and built a vast library and museum in Alexandria. The term *museum* was coined because the structure was dedicated to the seven muses, but it was really a research institute, the first state-run research institute in the world.

Ptolemy's successors treasured books, and had rather interesting means of obtaining them. Ptolemy II, desiring the first Greek translation of the Old Testament, "commissioned" the work by imprisoning seventy Jewish scholars in cells on the island of Pharos. Ptolemy III wrote to all the world's sovereigns requesting to borrow their books, then kept them. In the end, this aggressive approach to title acquisition worked: the library of Alexandria came to house a treasure of between 200,000 and 500,000 papyrus scrolls, depending on whose history you believe, representing most of the world's knowledge at that time.

The museum and library made Alexandria the unrivaled intellectual center of the world, a place where the greatest scholars of Alexander's former empire studied geometry and space. If *U.S. News & World Report* were to extend its survey of academic institutes to all of history, Alexandria would beat the Cambridge of Newton, the Göttingen of Gauss, and Einstein's Institute for Advanced Study at Princeton for the number one spot. Virtually all the great Greek mathematical and scientific thinkers to follow Euclid worked in this incredible library.

In 212 B.C., Eratosthenes of Cyrene, the chief librarian of Alexandria, a man who probably never ventured more than a few hundred miles in his life, became the first person in history to measure the circumference of the earth. His calculation caused a sensation among his fellow citizens, demonstrating just how small a piece of the planet was known to their civilization. Traders, explorers, and visionaries must have wondered wistfully about questions like "Is there intelligent life on the other side of the ocean?" A feat comparable to Eratosthenes' today would be to be the first to reveal that the universe didn't end at the outer reaches of our solar system.

Eratosthenes accomplished his insight about our planet without having to venture very far. Like Einstein, he succeeded by employing geometry. Eratosthenes noticed that at noon in the town of Syene (today called Aswan), during the summer solstice, a stick in the ground casts no shadow. To Eratosthenes, this meant that a stick stuck straight into the ground was parallel to the rays of the sun. Picturing the earth as a circle, a line drawn from its center through a dot on the circle representing Syene and out into space will be parallel to other lines representing the sun's rays. Now move along the circle of the earth's surface away from Syene, to Alexandria. Again draw a line from the center of the earth to a dot representing this city. This line is not parallel to the sun's rays. It intersects them at an angle, which is why one sees shadows.

The length of the shadow at Alexandria and a theorem in *Elements* about a line that crosses two parallel lines were enough to enable Eratosthenes to calculate the portion of the earth's circumference represented by the arc along the earth from Syene to Alexandria. He found that it represents one-fiftieth of the earth's circumference.

Employing perhaps the first ever graduate research assistant, Eratosthenes hired a nameless fellow to walk between the two cities, measuring their distance. He dutifully reported back that it came to about 500 miles. Multiplying that by 50, Eratosthenes determined the circumference of the earth to be about 25,000 miles, correct to within 4 percent, an astoundingly accurate answer that would certainly have earned him a Nobel Prize, and for the anonymous walker, perhaps, a tenured position at the library.

Eratosthenes was not the only Alexandrian of his time to make a major contribution to the understanding of the cosmos. Aristarchus of Samos, an astronomer working in Alexandria, used an ingenious and somewhat involved method combining trigonometry with a simple model of the heavens to calculate to a reasonable approximation the size of the moon, and its distance from the earth. Again, the Greeks obtained a new perspective on their place in the universe.

Another star attracted to Alexandria was Archimedes. Born in Syracuse, a city on the island of Sicily, he traveled to Alexandria in order to study at the royal school of mathematics. We may not know what genius first shaped stone or wood into a roundish form and astounded bewildered onlookers by demonstrating the first wheel, but we do know who discovered the principle of the lever: Archimedes. He also discovered the principle of buoyancy, and made many other contributions to physics and engineering. In mathematics, he carried the subject to a height not surpassed until the tools of symbolic algebra and analytic geometry were developed some eighteen centuries later.

One of Archimedes' mathematical achievements was to perfect a version of calculus not too distant from that of Newton and Leibniz. Considering the absence of Cartesian geometry, it was perhaps an even more impressive feat. He believed his crowning achievement to be his discovery via that method that the volume of a sphere inscribed in a cylinder (i.e., a sphere whose radius equals the radius and height of the cylinder) is two-thirds that of the cylinder. Archimedes was so proud of that discovery that he requested to have a diagram depicting it inscribed on his tomb.

When the Romans invaded Syracuse, Archimedes, then seventy-five, was murdered by a Roman soldier one day as he studied a geometric diagram he had drawn in the sand. His tomb was inscribed as he had wished. More than a hundred years later Cicero, the Roman orator, visited Syracuse and found Archimedes' tomb near one of its gates. Neglected, the tomb had become covered with thorns and briars. Cicero had it restored. Sadly, today it is nowhere to be found.

Astronomy, too, reached an apex in Alexandria, with the work of Hipparchus, in the second century B.C., and Claudius Ptolemy (not related to the kings of that name), in the second century A.D. Hipparchus observed the heavens for thirty-five years, then combined his observations with Babylonian data to work out a geometric model of our solar system in which the five known planets, the sun, and the moon all move in compound circular orbits about the earth. He was so successful in describing the motion of the sun and moon as seen from earth that he could predict lunar eclipses within a couple of hours. Ptolemy refined and extended this work in a book called the *Almagest,* which completed Plato's program of giving a rational explanation for the motion of bodies in the heavens, and dominated astronomical thought until Copernicus.

Ptolemy also wrote a book called *Geographia,* describing the terrestial universe. Cartography is a highly mathematical subject because maps are flat, yet the earth is nearly spherical, and a sphere cannot be mapped onto a part of a plane in a way which represents both areas and angles accurately. *Geographia* represented the start of serious mapmaking.

By the second century A.D., the fields of mathematics,

physics, cartography, and engineering had all made great strides. We knew that matter consisted of indivisible bits called atoms. We had invented logic and the proof, geometry and trigonometry, and a form of calculus. In astronomy and space, we knew that the world was very old, and that we lived on a sphere. We even knew the size of the sphere. We had begun to understand our place in the universe. We were poised to march on. Today we know that there are other solar systems only tens of light years away. Had the Golden Age continued unabated, we might by now have sent probes exploring them. We might have landed on the moon in the year 969 instead of 1969. We might have an understanding of space and life that is unimaginable to us today. Instead, events occurred that would delay the progress begun by the Greeks by a millennium.

There may have been more written about the causes of the intellectual decline of the Middle Ages than all the words in the Alexandrian library. There is no simple answer. The Ptolemaic dynasty declined in the two centuries preceding the birth of Christ. Ptolemy XII bequeathed his kingdom jointly to his son and daughter, who inherited it upon his death in 51 B.C. In 49 B.C. his son staged a coup against his sister, grabbing sole power for himself. His sister, not the type to acquiesce to such treatment, stole her way in to the visiting Roman emperor to plead her case (at the time, though technically independent of Rome, the Ptolemaic empire was already under Roman domination). So began the affair of Cleopatra and Julius Caesar. Eventually, Cleopatra claimed to have borne Caesar a son. He was a powerful ally for the Egyptian, but their alliance was doomed with Caesar himself. After twenty-three Roman senators pounced on their emperor and stabbed him to death on the Ides of March in 44 B.C., Caesar's great-nephew Octavian brought Alexandria and Egypt under Roman rule.

As Rome conquered Greece, the Romans became custodians to the Greek legacy. The inheritors of the Greek traditions conquered much of the world, and with it, faced many technical and engineering problems, yet their emperors did not support mathematics as did Alexander or the Ptolemys of Egypt, and their civilization did not produce mathematical minds such as Pythagoras, Euclid, and Archimedes. In the 1,100 years of their recorded existence, dating from 750 B.C., history does not mention one Roman theorem proved, nor even one Roman mathematician. To the Greeks, determining distances was a mathematical challenge involving congruent and similar triangles, and parallax, and geometry. In a Roman textbook, a word problem asked the reader to find a method for determining the width of a river *when the enemy occupies the other bank.* "The enemy"—a concept of questionable utility in mathematics, but a central one in Roman thinking.

In abstract mathematics, the Romans were ignorant, and proud of it. As Cicero said, "The Greeks held the geometer in the highest honor; accordingly, nothing made more brilliant progress among them than mathematics. But we have established as the limit of this art its usefulness in measuring and counting." Perhaps of the Romans we should say, "The Romans held the warrior in the highest honor; accordingly, nothing made more brilliant progress among them than raping and pillaging. But we have established as the limit of this art its usefulness in conquering the world."

It is not that the Romans weren't literate. They were. They even wrote their own Latin technical books, but these were bastardized works adapted from their knowledge of the Greeks. For example, the principal translator of Euclid into Latin was a Roman senator from an old established family, Anicius Manlius Severinus Boethius, a kind of *Reader's Digest* editor of Roman times. Boethius abridged Euclid's works, creating the kind of treatment suitable for students preparing for a multiple-choice test. Today his translations might be entitled *Euclid for Dummies* or sold in TV ads imploring, "Call 1-800-NOPROOFS," but in Boethius' time, his were the authoritative works.

Boethius gave only definitions and theorems, and apparently also felt free to substitute approximations for exact results. And that was on a good day. In other cases, he just plain got it wrong. For his misrepresentation of Greek ideas, Boethius was not flayed, crucified, burned at the stake, or subjected to any of other popular punishments for intellectuals of the medieval period. His downfall came because he also got involved in politics. In 524, he was beheaded for having "treasonable contacts" with the Eastern Roman Empire. He should have stuck to bastardizing mathematics.

Another book typical of the recidivism of the period was written by a well-traveled merchant of Alexandria. "The Earth," this Roman wrote, "is flat. The inhabited portion has the shape of a rectangle whose length is double its breadth. . . . In the north is a conical-shaped mountain around which the sun and moon revolve." His book, *Topographia Christiana,* was based not on reason or observation, but on the Scriptures. A good book to read between sips of tasty lead-laden Roman wine, *Topographia* remained on the bestseller list until the twelfth century, long after the Romans were history.

The last great scholar to work at the library in Alexandria was named Hypatia, the first great woman scholar whose story has been passed down by history. She was born in Alexandria around A.D. 370, the daughter of a famous mathematician and philosopher named Theon. Theon taught his daughter mathematics. She became his closest collaborator, but eventually completely eclipsed him. One of her former students, Damascius, who later turned out to be a harsh critic, wrote that she was "by nature more refined and talented than her father." Her fate and its larger meaning have been much discussed over the centuries, by such diverse authors as Voltaire, and Edward Gibbon in *The Decline and Fall of the Roman Empire*.

Around the turn of the fifth century, Alexandria was one of the greatest strongholds of Christianity. This caused a huge struggle for influence between the representatives of the church and the state. It was a time in Alexandria of many social disturbances and conflicts between Christians and non-Christians, such as Greek Neoplatonists and Jews. In 391, a Christian mob attacked and burned most of the library at Alexandria.

On October 15, 412, the Christian archbishop of Alexandria died. He was succeeded by his nephew, a man named Cyril, often described as power-hungry and generally unpopular. The secular authority at the time was a fellow named Orestes, the prefect of Alexandria and civil governor of Egypt in the years 412–15.

Hypatia traced her intellectual heritage back to Plato and Pythagoras, not through the Christian Church. Some say she even chose to study in Athens, where she earned the laurel wreath, bestowed upon only the best of Athens's pupils, and on her return to have worn this wreath whenever she appeared in public. She reportedly wrote significant commentaries on two famous Greek works, Diophantus' *Arithmetica,* and Apollonius' *Conic Sections,* two works that are still read today.

Said to have been a great beauty and a charismatic lecturer, Hypatia held widely attended public lectures on Plato and Aristotle. According to Damascius, the whole city "doted on her and worshipped her." At the end of each day, she would mount her chariot and ride to her lecture hall at the academy, a highly adorned room, with swinging lamps of perfumed oil and a huge rotunda handpainted by a Greek artist. Hypatia, wearing a white robe and her ever present laurel, would face the large crowd and transfix them with her eloquent Greek. She attracted students from Rome, Athens, and other great cities of the empire. One who attended her lectures was the Roman prefect Orestes.

Orestes became Hypatia's friend and confidant. They met frequently, discussing not only her lectures but also municipal and political issues. This put her squarely on one side of Orestes' struggle with Cyril. She must have appeared to Cyril as a great threat, for her disciples held high positions, both within Alexandria and on the outside. Hypatia had the courage to continue her lectures, though Cyril and his follow-ers spread rumors that she was a witch who practiced black magic and cast Satanic spells on the people of the city.

There are several versions, most of them similar, about what happened next. One morning during Lent in 415, Hypa-tia climbed into her chariot, some say outside her residence, some say on a street intending to ride home. Several hundred of Cyril's stooges, Christian monks from a desert monastery, swooped upon her, beat her, and dragged her to a church. In-side the church they stripped her naked and peeled away her flesh with either sharpened tiles or broken bits of pottery. Af-terward, they ripped apart her limbs and burned her remains. According to one account, parts of her body were scattered all over the city.

Hypatia's works were all destroyed. Not long afterward, so were the remnants of the library. Orestes left Alexandria, probably recalled, and in historical documents was never heard from again. Future imperial officials accommodated Cyril with the influence he had sought. He was eventually canonized.

A recent historical study estimates, throughout history, that there has been an average of one memorable mathematician for every three million people. Today, a work of research is widely accessible throughout the world. In the fourth century, when scrolls had to be painstakingly hand-copied with primitive pens, a lost book put the work itself on the endangered species list. We cannot know what great treasures of Babylonian and Greek mathematics were lost forever with the burning of the library's more than 200,000 scrolls. We do know that the library contained over a hundred plays by Sophocles, and, of these, only seven survive today. Hypatia was the embodiment of Greek science and rationalism. With her death, came the death of Greek culture.

With the fall of Rome around A.D. 476, Europe inherited great stone temples, theaters and mansions, and modern municipal services like street lights, running hot water, and sewage systems, but little in the way of intellectual achievement. By 800, only fragments of a Latin translation of Euclid's *Elements* still existed. Grafted onto a collection of surveying texts, they contained only formulae, freely utilizing approximations and making no attempt at derivation. The Greek tradition of abstraction and proof appeared lost. As the brilliant Islamic civilization thrived, Europe slid into deep intellectual decline. Hence the name given to this time in Europe: the Dark Ages.

Eventually, Greek thought would be resurrected. Books like *Topographia* fell from favor, and Boethius' works were replaced by more faithful translations. In this late medieval period, a group of philosophers created an atmosphere of reason that allowed the great sixteenth-century mathematicians like Fermat, Leibniz, and Newton to flourish. One of these thinkers was at the center of the next revolution in geometry and our understanding of space. His name was René Descartes.

II

THE STORY

OF DESCARTES

Where are you in space?
How mathematicians
discovered the simple
principles of graphs and
coordinates that led to
epic breakthroughs in
philosophy and science.

7. The Revolution in Place

OW DO you know where you are? After the realization that space itself exists, this is perhaps the next natural question. It may seem that the answer is provided by cartography, the study of maps. But cartography is only the beginning. A proper theory of place leads to ideas far deeper than simple statements like "To find Kalamazoo, look in F3."

There is more to location than naming a spot. Imagine an alien emissary landing on earth, a stringy bubble-headed creature living on oxygen, or perhaps a hairy, apelike individual partial to nitrous oxide. If we wished to communicate, it would be nice if the alien had brought a dictionary. But would that be enough? If your idea of good communication is "Me Tarzan you Jane," it might be, but for an exchange of intergalactic ideas we'd also have to learn each other's grammar. In mathematics, too, the "dictionary"—a system of naming the points in the plane, in space, or on the globe—is just a beginning. The real power of a theory of location resides in the ability to relate different locations, paths, and shapes to each other, and to manipulate them employing equations—in the unification of geometry and algebra.

Today, as one old textbook on the subject states, "With relatively little effort the student may now reach out and grasp these tools." It is hard to imagine what yet greater theories the great astronomer/physicists Kepler and Galileo could have created had the tools of coordinate geometry been familiar to them, but they had to do without. With this knowledge, their successors Newton and Leibniz created calculus and the modern age of physics. Had geometry and algebra remained unrelated, few of the advances of modern physics and engineering would have been possible.

Like the revolution of proof, the first signpost along the way to the revolution of place came in pre-Greek times, with the invention of maps. Though the Greeks added their particular genius, the end of their civilization left the subject unfinished, and the power unleashed. The next step along the way was the invention of the graph, but this awaited the revival of the intellectual tradition following the Dark Ages. In the end, this revolution trailed by a dozen centuries the last great Greek mathematicians and cartographers.

8. The Origin of Latitude and Longitude

O ONE knows who made the first maps, or when, or why. We do know that some of the earliest known maps were created for the same reason that the Egyptians invented geometry. These maps, simple clay tablets dating back to 2300 B.C., had inscribed upon them, not topographic keys or religious ornamentation, but notes regarding property tax. By 2000 B.C. real estate maps describing data such as property outlines and owners were common in Egypt and Babylon. One imagines a bejeweled Mesopotamian woman, her expression a bit strained from the weight of the clay tablet in her grasp, pointing to a spot upon it and solemnly chanting in her ancient language: "Location, location, location."

As more and more brave souls began to explore the seven seas, a more vital purpose dominated the creation of maps. As recently as 1915, when Sir Ernest Shackleton's ship, the *Endurance,* became trapped and broke up in the Antarctic winter, the greatest danger for the crew came not from the winds of nearly 200 miles per hour nor the temperatures reaching −100 degrees Fahrenheit, but from the problem of finding their way back. Throughout history, it has been this way. The most vital challenge facing seafarers and explorers on the open ocean has been the challenge of not getting lost. Suppose you are stranded with absolutely no information about where you are. You have with you no instruments of navigation, but a two-way radio you can use to call in for help. How could you tell your rescuers where you are?

The two coordinates that we use to describe your position today on the surface of the earth are latitude and longitude. To picture these, place in your mental toolbox three points, two lines, and a globe. Take the globe from your toolbox and imagine it floating in space. This, of course, represents the earth. Next, place your three points as follows: stick one on the earth's North Pole, one at its center, and the third at some location on the surface. Use your first line to connect the North Pole with the earth's center. This is the earth's axis of rotation. Use the other line to connect the earth's center to the point on the surface. This line will then make a certain angle with the earth's axis. That angle, apart from conventions, determines your latitude.

The original idea of latitude came from an ancient meteorologist named Aristotle. After studying how placement on the earth affects climate, he proposed dividing the globe into five climatic zones delineated by north/south location. These zones were eventually included in maps, separated by lines of constant latitude. As Aristotle's theory suggests, you can determine your latitude, at least on average, by climate—the earth is coldest at the poles, and warms as you move toward the equator. Of course, on any particular day Stockholm may be warmer than Barcelona, so unless you are willing to sit around making measurements over a period of time, this method is not really useful. A better way to determine latitude is to examine the stars. This is especially simple if you find a star along the earth's axis. There is such a star in the northern hemisphere, Polaris, the "polestar."

Polaris hasn't always been the polestar, because the earth's axis isn't precisely fixed with respect to the stars. It precesses, tracing out a narrow cone with a period of 26,000 years. Some of the great pyramids of ancient Egypt have passages aligned in the direction passed over by α-Draconis: when they were built, α-Draconis was the polestar. The ancient Greeks had it harder—in their day there was no true polestar. Some 10,000 years from now, the polestar (in the north) will be easy to find. It will be Vega, the brightest star in the northern sky.

If you can simultaneously sight Polaris and the horizon to the north, then simple geometry shows that the angle between the lines from you to these points is approximately your latitude. The relation is only approximate because it assumes that Polaris lies exactly along the earth's axis, and that the radius of the earth is negligible compared with the distance to Polaris; both good, but not perfectly accurate, assumptions. In 1700, Isaac Newton invented the sextant, a device designed to ease the process of the sightings and measuring latitude in this way. The stranded traveler could do it the old-fashioned way, though, employing two sticks as a protractor.

Determining your longitude is harder. Add to your mind's image another sphere, much larger than the earth, with the earth at its core. On this sphere imagine a map of the stars. If the earth didn't rotate, you could measure your longitude with reference to this map. The effect of the earth's rotation, though, is that the star map you see one moment will be the star map a person a little to your west sees a moment later. To be precise, since the earth rotates through 360 degrees in 24 hours, an observer 15 degrees to your west sees the same view as you an hour later. At the equator, this difference corresponds to roughly 1,000 miles. Comparing two snapshots of the stars taken at the same latitude, but without time stamps, gives no information about your longitude. On the other hand, if you compare snapshots taken at the same latitude and at the same time of night, you can determine from them your difference in longitude. But for that, you need a clock.

Not until the eighteenth century were clocks made that could withstand the motion, the changes in temperature, and the salty moisture that come with ships at sea and still be accurate enough to be useful for determining longitude over vast stretches of ocean. The accuracy requirement was not trivial: an error of just three seconds per day over a voyage of six weeks corresponds to an error in longitude of over half a degree. Until the nineteenth century, there were also many different conventions employed for defining longitude. Finally, in October 1884, a single meridian was agreed upon by the world at large as the "zero" of longitudes from which longitude differences would be measured, the "prime meridian," passing through the Royal Observatory at Greenwich, outside London.

The first great world map created by the Greeks was drawn by Thales' student Anaximander around 550 B.C. His map divided the world into two parts, Europe and Asia. The latter in his map included North Africa. By 330 B.C., the Greeks were even putting maps on some of their coins; one of them included elevations and is said to be "the first physical relief map known."

The Pythagoreans, in addition to all their other momentous contributions, appear to have been the first to have proposed that the earth is a sphere. This concept is, of course, vital to accurate mapmaking, and, fortunately, had powerful proponents in Plato and Aristotle even before Eratosthanes more or less proved it by applying a spherical model to measure the earth's circumference. After Aristotle proposed his idea of dividing the world into climatic zones, Hipparchus invented the idea of spacing these at equal intervals, and adding north-south lines at right angles. By the time of Ptolemy, about five centuries after Plato and Aristotle, and four after Eratosthanes, the names "latitude" and "longitude" had been given to these lines.

In his *Geographia,* Ptolemy appears to have used a method similar to stereographic projection to represent the earth on a flat surface. To locate positions, he employed latitude and longitude as coordinates. He assigned them to every place on earth with which he was familiar—8,000 in all. His book also contained instructions for mapmaking. *Geographia* was a standard reference for hundreds of years. Cartography, like geometry, was poised for entry into the modern age. But like geometry, the subject made no progress under Roman rule.

The Romans produced maps, but like the geometry problem that focused on enemy troops across the river, these efforts were focused on purely practical, often military problems. When the Christian mob sacked the library at Alexandria, *Geographia* disappeared along with the mathematical works of the Greeks. When Rome fell, the new age found civilization as much in the dark about describing place in space as it was about the theorems and relations among spatial objects. Geometry and cartography would eventually be reborn and revolutionized by a new theory of place. Before this could happen, a larger task had to be accomplished: the revival of the intellectual traditions of Western civilization.

9. The Legacy of the Rotten Romans

HE TIME was late in the eighth century. The great works and traditions of the Greeks were lost and forgotten; the clock and the compass were as far in the future as is to us the *Starship Enterprise*. As they lay in their beds or on the hard floor, shivering or sweating, waiting for sleep, the inhabitants of the times did not mutter to themselves, "Unless I revive the pursuit of knowledge, this period of intellectual decay and stagnation will not improve for nearly a thousand years." Yet in this age one powerful man did recognize the need for more learning, and take the steps that would eventually lead to the rebirth of an intellectual tradition in Europe.

Genetically, Charles the Great, or Charlemagne, might have seemed a long shot. Measured by his skeleton after death, he proved to be 6 feet 4, a giant for his day. His father, whom Pope Stephen elevated to King Pepin I in 754, was a diminutive man previously known as Pepin the Short. Charlemagne's stature came presumably from his mother, Queen Bertha. Her skeleton was not measured after death, yet her nickname hints at her build: she was called *grand pied,* or "big foot."

Charlemagne was powerful in all respects: physique, intellect, and perhaps most importantly, the size of his army.

Charlemagne had a "knock a wall down here, move a wall there" philosophy of kingdom ownership, which he applied to the map of Europe. He enlarged the territory of his Frankish kingdom by knocking down the borders of his neighbors, the Lombards, the Bavarians, and the Saxons. He became the dominant force in Europe and imposed Roman Catholicism everywhere he ventured. If this was all he had done, he might have been just another king with a hobby of world domination. But Charlemagne was a patron of education reminiscent of Alexander. He realized he had inherited a dearth of teachers, so he invited the most prominent educators in his kingdom and beyond to his court in Aachen, where he set up the Palace School. In this he took a particular interest, once personally whipping a boy who had made a mistake in Latin. We do not know whether Charlemagne also practiced self-flagellation, but he himself was illiterate, though he made several failed attempts to learn to write. (The whipping may not seem as harsh in light of other penalties of the day, such as the penalty for eating meat on Friday: death.)

Under Charlemagne, the Christian Church, requiring armies of literate monks to do its bidding, became the driving force of scholarship. Church schools were organized, attached to cathedrals or monasteries, with teachers usually provided by church orders such as the Dominicans and Franciscans. They trained priests, prepared a literate aristocracy, and restored a respect for the classics. Scribes set to work producing numerous copies of manuscripts in their archives—textbooks, encyclopedias, anthologies. To increase efficiency, the monks developed a new style of handwriting called Carolingian miniscule that is still the basis for our writing in the Latin alphabet today. Charlemagne was equally proactive in taking care of himself. It is an emblem of the times that, in his quest for long life, he did not employ a bevy of alchemists or gather around him an academy of doctors. Instead, he invented a kind of theological factory, an industry of clergy devoted to his health. In one monastery alone, Charlemagne had 300 monks and 100 clerks praying continuously for him, in three shifts, around the clock. He died anyway. The year was 814.

Charlemagne's revival produced little in the way of original work. Upon his demise, his kingdom contracted and his successors did not extend his cultural Renaissance. Still, the level of literacy never again fell to that of pre-Carolingian (that is, pre-Charlemagne) times. The church schools he fostered, though hardly insulated bastions of independent discourse, spread like wildflowers and eventually turned into the universities of Europe, beginning, according to most historians, with the University of Bologna in 1088. It was these that would eventually allow Europe to reemerge as an intellectual power, and France, especially, as a center of mathematics. By the turn of the millennium, the Dark Ages had ended. What we call the Middle Ages would continue for about another 500 years.

Through trade, travel, and the Crusades, Europeans eventually came into contact with Arabs of the Mediterranean and the Near East, and the Byzantines of the Eastern Roman Empire. In the case of the Crusades, "contact" with the Europeans was about as desirable as contact with the Martians in *War of the Worlds*. But even as the Europeans plundered Arab lands and mercilessly slaughtered the Muslim and Jewish infidels, they also coveted their wisdom. While mathematics and science in the West had withered, the Islamic world had retained faithful versions of many Greek works, including Euclid and Ptolemy. Though they, too, made little progress in abstract mathematics, they made significant advances in calculational methods. Propelled by the religious needs of time and calendar, they had developed all six trigonometric functions, and perfected the astrolabe, a hand-held instrument allowing accurate observation of the altitude of a star or planet.

Church and secular leaders supported scholars in their hunt for the knowledge of their foes, and also for lost Greek intellectual treasures, in the original or Arabic translation. Early in the twelfth century, the Englishman Adelard of Bath traveled to Syria disguised as a Mohammedan student. He later translated Euclid's *Elements* into Latin, this time with proofs. A century later, Leonardo of Pisa, also known as Fibonacci, brought from North Africa the idea of zero, and the Hindu-Arabic number system we use today. The influx of ancient Greek knowledge fed the new universities.

The stage was set for another Golden Age akin to that of the Greeks. The comparison wasn't lost on those living at the time. An English monk named Bartholomew wrote, "Just as the town of Athens in older times was the mother of the liberal arts and letters, the nurse of philosophers and all manner of science, such is Paris in our day. . . ." Unfortunately, practical matters got in the way.

In mathematician Andrew Wiles's recent (successful) quest to prove Fermat's last theorem, he relied on his academic lifestyle of quiet contemplation. Wiles was working some 350 years after Fermat. An equal number of years before Fermat was the time of peak achievement in medieval mathematics. The life of a medieval professor included no cookie-catered seminars, no days of serene concentration punctuated by a stroll across campus, no great mathematicians flying in for a visit which includes a nice faculty dinner at the local Chinese restaurant. Everyone knows that Europe in the Middle Ages was no Garden of Eden. But if you are caught in a cheap science fiction movie and the mad scientist gives the dial on his time machine a random spin, you'd better pray it doesn't land on the thirteenth or fourteenth centuries.

The medieval mathematician faced steaming summers, freezing winters, and after sunset, buildings that were poorly heated and virtually unlit. On the street, wild pigs ran free as scavengers, the blood of slaughtered animals streamed from butcher shops, and discarded chicken heads flew from the poultry store entrance. Only the large cities had sewage systems. Even King Louis IX of France was once splashed by contents, unmentionable in these pages, tossed onto the street from above.

The weather gods, too, were in a foul mood. Europe at the time was at the start of a wet and cold period so distinctly miserable that today it is called the little ice age. In the Alps, glaciers advanced for the first time since the eighth century. In Scandinavia, ice floes blocked the North Atlantic shipping lanes. Crops failed. Agricultural productivity plummeted. Famine was widespread. In England, common people ate dogs, cats, and other novel dishes described in one account only as "unclean things." The aristocracy suffered correspondingly: they were reduced to eating their horses. According to one account of a famine in the Rhineland, troops had to be posted at the gallows in Mainz, Cologne, and Strasbourg to defend against ravenous citizens who were cutting down and eating the corpses.

In October 1347, a fleet coming from the Orient landed in northeast Sicily. Unfortunately for the continent of Europe, they had known enough geometry to find their way to port. It was their medical knowledge that was inadequate. All aboard were dead or dying. The crew was quarantined. The rats scampered away, carrying the Black Death to the shores of Europe. By 1351, up to a half of Europe's population had died. The Florentine historian Giovanni Villani wrote: "It was a disease in which there appeared certain swellings in the groin and under the armpit, and the victims spat blood, and in three days they were dead. . . . And many lands and cities were made desolate. And the plague lasted till _____." Villani left a blank at the end of his report, meaning to fill in the year in which the plague finally ended. If that seems like a good way to jinx yourself, it was: he died of the plague in 1348.

College provided no haven from these miserable conditions. The concept of a college campus did not yet exist. Typically, a university had no buildings at all. Students lived in cooperative housing. Professors lectured in rented rooms, rooming houses, churches, even brothels. The classrooms, like the dwellings, were poorly lit and heated. Some universities employed a system that sounds, well, medieval: professors were paid directly by the students. At Bologna, students hired and fired professors, fined them for unexcused absence or tardiness, or for not answering difficult questions. If the lecture was not interesting, going too slow, too fast, or simply not loud enough, they would jeer or throw things. In Leipzig, the university eventually found it necessary to promulgate a rule against throwing stones at professors. As late as 1495, a German statute explicitly forbade anyone associated with the university from drenching freshmen with urine. In numerous cities, students rioted and fought with the townspeople. Across Europe, it was the fate of college professors to deal with behavior that would make *Animal House* seem like an instructional video for good manners.

The science of the day was a hodgepodge of ancient knowledge intertwined with religion, superstition, and the supernatural. Belief in astrology and miracles was common. Even great scholars like St. Thomas Aquinas accepted without question the existence of witches. In Sicily, the emperor Frederick II founded the University of Naples in 1224, the first university founded and run by laymen. Unhampered by the annoying concept of ethics, Frederick indulged his love for science with the occasional experiment on humans. One time he fed two lucky prisoners the same huge and lavish meal. He sent one of the happy men to bed, and the other on a grueling hunt. Afterwards, he cut them both open to see who had better digested the meal. (Couch potatoes will be pleased to note it was the man who slept.)

The concept of time was vague. Until the fourteenth century, no one knew with any precision what time it was. Daylight, divided into twelve equal intervals based upon the sun's journey overhead, consisted of hours whose length varied with the season. In London, at a latitude of 51½ degrees north, where the period from sunrise to sunset is more then twice as long in June as December, the medieval hour varied between approximately 38 and 82 of today's minutes. The first clock recorded as striking equal hours didn't come until the 1330s, in the church of St. Gothard in Milan. In Paris, a public clock didn't come until 1370, on one of the towers of the Royal Palace. (It still exists today, on the corner of the boulevard du Palais and the quai de l'Horloge.)

No technology existed to accurately measure short time intervals. Rates of change such as speeds could only be quantified roughly. Fundamental units, like the second, were rarely used in medieval philosophy. Instead, continuous quantities were vaguely described as having a certain "degree" of magnitude, or else were assigned a size only through pairwise comparison. For instance, a particular chunk of silver might be said to weigh one-third as much as a plucked chicken, or twice as much as a mouse. The awkwardness of this system was aggravated by the fact that the principal medieval authority on numerical ratios was a book called *Arithmetica* by Boethius, and Boethius did not use fractions to describe ratios. For medieval scholars, the ratios that described quantities were not regarded as numbers, and could not be manipulated using arithmetic the way numbers can.

Cartography, too, was primitive. Maps in medieval Europe were not meant to portray exact geometrical and spatial relationships. They were not constructed on geometric principles, nor was there much notion of scale. Instead, they were usually symbolic, historical, decorative, or religious.

With all this to hamper progress of the mind, the main impediment was a more direct constraint: medieval scholars were required by the Catholic Church to take it for granted that the Bible was literally true. The church taught that every mouse, every pineapple, every housefly served a purpose in

God's scheme, and that this scheme could be understood only from the Scriptures. To propose otherwise was dangerous.

The church had good reason to fear the rebirth of reason. If the Bible is divinely inspired, then its authority, regarding both nature and morality, rests on the Bible's absolute acceptance. Yet the Bible's description of nature often clashed with concepts of nature derived from observation or mathematical reasoning. In nurturing the universities, the church therefore unwittingly contributed to the decline of its own authority in both nature and morality. But the church did not stand on the sidelines and see its primacy undermined.

● ■ ▲

The main movement in natural philosophy in the late Middle Ages was that of the Scholastics, centered in the new universities, especially at Oxford and Paris. Seeking an intellectual armistice, the Scholastics spent much of their energy attempting to reconcile their physical theories with their religion. A central question in their philosophy became not the nature of the universe, but the "meta-question" of whether the knowledge given in the Bible could also be derived or explained through the application of reason.

The first great Scholastic argued for logical discussion as a method of deciding truth. He was a twelfth century Parisian, Peter Abelard. In medieval France, his was a dangerous stance to take. Abelard was excommunicated; his books were burned. The most famous Scholastic, St. Thomas Aquinas, was also a proponent of reason, but he was one the church could endorse. Aquinas approached knowledge in the manner of the True Believer, or at least as someone not wishing to see his books as the fuel warming huddled monks on a cold winter night. Rather than setting out to follow an argument wherever it may lead, Aquinas began by accepting truth as declared in the Catholic faith, and then sought to prove it.

Though Aquinas was not condemned by the church, he was roundly attacked by a contemporary Scholastic, Roger Bacon. Bacon was one of the first natural philosophers to place enormous value on experiment. If Abelard got into trouble for emphasizing reason over the Scriptures, Bacon's heresy was placing emphasis on truth derived from observation of the physical world. In 1278 he was condemned to prison, where he remained for fourteen years. He died shortly after his release.

William of Occam, a Franciscan at Oxford and later at Paris, is famous for "Occam's razor," the aesthetic that still holds in physical science today. Stated simply, it is this: One should strive to create theories based on as few ad hoc assumptions as possible. One of the motivations for string theory, for example, is to derive fundamental constants such as the charge of the electron, the number (and type) of "elementary particles" that exist, even the number of dimensions of space. In previous theories such information was always axiomatic—included in the construction of the theory, not derived from it. In mathematics, a similar aesthetic applies: when creating a theory of geometry, for instance, one seeks to use the minimal number of axioms necessary.

Occam became involved in a quarrel between the Franciscan order and Pope John XXII, and was excommunicated. He escaped, took refuge with Emperor Louis, and settled in Munich. He died in 1349 at height of the plague.

Of Abelard, Aquinas, Bacon, and Occam, only Aquinas escaped unscathed. Abelard, in addition to being excommunicated, was castrated for having beliefs on marriage not in alignment with his girlfriend's uncle, who happened to be a canon in the Catholic Church.

The Scholastics contributed a great deal to the intellectual revival of the Western world. One of their beneficiaries was an obscure French cleric from the village of Allemagne, near

Caen. From the point of view of mathematics, his was the most promising work. In modern books on astronomy and mathematics, this man, who became the bishop of Lisieux, is scarcely mentioned. At his University of Paris, he is not highly honored. In the Cathedral of Notre-Dame, the memorial candles his brother Henri commissioned have long since ceased shedding light. On earth, his memorials are few, but it is fitting that on a journey to the moon one of the features you'll find is a crater named in his honor, the lunar feature called Oresme.

10. The Discreet Charm of the Graph

EEP IN the Amazon rainforest, a tough, river-wise woman boats down tributaries home to blood-sucking fishes and swarming mosquitoes to stop at forest huts rarely graced by anyone outside their few isolated inhabitants. She is not a character from the Middle Ages. She lives today. Who is she? A doctor perhaps? A foreign aid worker? Not even warm. She is peddling creams, perfumes, and cosmetics for the Avon company.

Back at the New York headquarters, suited executives analyze their worldwide war against dry skin employing techniques invented by a man to whom one can safely say they have never given any thought. International in blue, domestic in red, one may imagine, graphs compare the year-by-year erection of Avon's profits in each sector. Their annual report analyzes the company's cumulative return, net sales, business unit operating profit, and pages of other data utilizing all sorts of fancy graphs, bar graphs, and pie charts.

A merchant presenting data in this way in the Middle Ages would have been greeted by a blank stare. What is the meaning of these colorful geometric figures, and why do they appear on the same document with all those Roman numerals? Macaroni and cheese had been invented (a fourteenth-century English recipe survives), but not the idea of marrying numbers and geometric figures. Today, graphical representation of knowledge is so familiar that we hardly think of it as a mathematical device: even the most math phobic executive at Avon could tell that an upward slanted line on the profits graph is a happy thing. But upward or downward, the invention of the graph was a vital step on the path to a theory of place.

The marriage of numbers and geometry is one concept the Greeks got wrong, a spot on the road where philosophy got in the way. Today, every schoolchild studies the number line, roughly speaking, a line endowed by an ordered correspondence between its points and the positive and negative whole numbers, as well as all the fractions and other numbers in between. These "other numbers" are the irrationals, numbers which are neither whole numbers nor fractions, but which, as Pythagoras refused to admit, seem to arise anyway. The number line must include them: without irrationals, it has an infinity of holes.

As we have seen, Pythagoras discovered that a square whose side is one unit long has a diagonal whose length, the square root of 2, is irrational. If this diagonal were laid along the number line with one end at zero, we could use its other end to mark out the point corresponding to the irrational number the square root of 2. When Pythagoras banned discussion of the irrationals because they didn't fit into his idea that all numbers were either whole or fraction, he realized that he also had to ban the association of line with number. In doing so he swept his problem under the rug but he also forbade one of the most fruitful concepts in the history of human thought. Nobody is perfect.

One of the few advantages of the loss of the Greek works was the faded influence of Pythagoras' views on the irrationals. The theory of the irrational numbers wasn't put on firm ground until Georg Cantor's and his contemporary Richard Dedekind's work in the late nineteenth century. Still, from the Middle Ages until then, most mathematicians and scientists ignored the fact that the irrationals seemed not to exist, and happily, if awkwardly, used them anyway. Apparently the reward of getting the right answer outweighed the unpleasantness of working with numbers that didn't exist.

Today the use of "illegal" mathematics is quite common in science, especially physics. The theory of quantum mechanics, for example, as worked out in the 1920s and 1930s, relied heavily on an entity invented by the English physicist Paul Dirac called the *delta function*. According to the mathematics of the time, the delta function was, simply, equal to zero. According to Dirac, the delta function was zero everywhere except at one point, where its value was infinite, and, when used in conjunction with certain operations of calculus, yielded answers that were both finite and (typically) non-zero. Later, the French mathematician Laurent Schwartz was able to show how the rules of mathematics could be redefined to allow for the delta function, and a whole new discipline of mathematics was born. The quantum field theories of modern physics may also be illegal theories of this sort—at least no one has yet successfully shown that, mathematically speaking, such theories legally exist.

Medieval philosophers were quite good at saying one thing and writing another, or even at writing one thing and also writing its contradiction—whatever it took to save their skin. So around the middle of the fourteenth century, Nicole d'Oresme, later bishop of Lisieux, didn't seem worried by any contradiction caused by the irrationals when he invented the graph. Oresme implicitly ignored the question of whether whole numbers and fractions are enough to fill out the graph's baseline. He focused on how his new pictures could be used to analyze quantitative relationships.

On one level a graph is a picture of a function, representing how one quantity varies as another is varied. The profits of Avon's Third World operations versus time, your calories burned versus distance walked, the day's high temperature versus geographical location, are all examples of functions. Each can be better understood employing graphs. The graph in the last example has a special name that hints of a deeper connection. It is a map, a weather map.

Any map is a kind of graph. For instance, a "normal" geopolitical map graphs the name of cities and countries, and perhaps some other data, versus geographical location. The Greeks and others had been making use of this kind of graph, the map, for thousands of years without realizing it. It is not clear whether Oresme realized it either, but he did touch on a central question: Does the curve or other shape formed by the graph of a collection of data, or of a function, have any geographical or geometric significance?

If we graph elevation versus location, we get the familiar topographical map whose connection to real geography is obvious. A duck-shaped mountain on a relief map is represented by the shape of a duck. But if we graph the weather with respect to location, we also obtain a surface, not literally the shape of the weather, but a geometric shape whose significance we can study. By relating functions to geometry in this way, we achieve a correspondence between types of function and types of shape. The study of lines and surfaces thus becomes the study of particular functions, and vice versa; we achieve a unification of geometry and number. It is this step that gives Oresme's invention of graphs its significance in mathematics.

The power of graphs in helping the non-mathematician analyze patterns of data stems from this same connection of data to geometry. The human mind easily recognizes certain simple shapes—lines and circles, for instance. When looking at a collection of points, our mind tries to fit them into one of these familiar patterns. As a result, we notice geometric patterns when data is graphed that we might easily overlook when staring at a table of numbers. The art of graphing is analyzed from this perspective in Edward Tufte's classic book *The Visual Display of Quantitative Information.*

Consider the three fairly dull-looking columns of numbers below.

Time	Alexei's Data	Nicolai's Data	Mom's Data
0	0.2	4.0	9.0
1	1.6	5.0	8.9
2	5.0	6.2	8.7
3	4.4	7.2	8.3
4	5.8	8.1	8.1
5	7.2	8.5	7.6
6	8.8	8.3	6.6
7	10.5	7.8	5.6
8	11.8	6.6	4.1
9	13.3	5.6	0.1
10	14.8	4.0	—

Each column is meant to represent a set of measurements, so each number includes experimental error. We'll call the first set of data Alexei's, as if a student named Alexei had taken the measurements, and the other two sets Nicolai's and Mom's. The question is, in each case, if we consider the data points as a function of time, is there a pattern, and if so, what is it?

The patterns, difficult to determine from the raw numbers, are immediately obvious if we graph the data. From Alexei's graph, it is easy to see that the data form a line, except for the aberrant point at time 2, when Alexei either sneezed or was distracted by a friend with a video game. In Nicolai's graph, the relationship between the data and the time is a well-known type of curve called a *parabola,* which represents, for

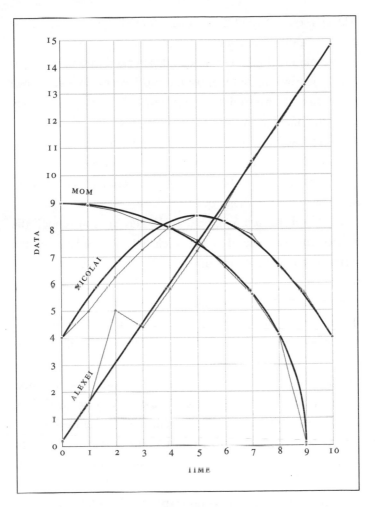

DATA TAKING SHAPE

instance, the energy of a spring plotted against its extension, or the height of a cannonball versus its distance traveled. Mathematically, this shape is described by a function where the measured data increase with the square of time (or distance). Mom's graph shows the top right quadrant of a circle, one of the most common shapes in our lives, and, like

Alexei's, one of Euclid's fundamental shapes. Yet from the numbers alone, that is far from obvious.

Oresme employed his new and powerful geometric technique to prove one of the most famous laws of physics known at the time, the Merton rule. Between 1325 and 1359, a group of mathematicians at Merton College, Oxford, had proposed a conceptual framework for describing motion quantitatively. In ancient discussions, distance and time had been considered quantities that could be described numerically, but "quickness" or "speed" was not quantified.

A central theorem conjectured by the Merton school, the Merton rule was a kind of yardstick for the race between the tortoise and the hare. Imagine a fictitious tortoise racing for a minute at, say, one mile per hour. Imagine the hare, starting even more slowly, but accelerating at a constant rate until at the end of the minute she is hopping much faster than her steady adversary. According to the Merton rule, if after the minute of constant acceleration, she is hopping at twice the speed of the tortoise, they will have gone the same distance. If she has reached a higher speed, she will be ahead, and if she has not yet achieved twice his speed, she will lag behind.

In more academic terms, the rule states that the distance traversed by an object accelerating uniformly from rest is equal to the distance traversed by an object moving for the same time at half the maximum speed. Given the murky understanding of place, time, and speed, and the inadequacy of means for measurement, the Merton rule is impressive. But without the tools of calculus or algebra, the Merton scholars could not prove their speculation.

Oresme proved the rule geometrically, employing his method of the graph. He began by placing time along the horizontal axis, and velocity along the vertical. With this technique, uniform velocity was to be represented by a horizontal line, constant acceleration by a line rising at some angle. Oresme realized that the area under these curves—respectively, a rectangle and a triangle—would represent the distance traversed.

The distance traveled by the object accelerating uniformly in the Merton rule is then given by the area of a right triangle whose base is proportional to the time traveled and whose height represents its maximum speed. The distance traveled by the object moving with constant speed is given by the area of a rectangle with the same base as the triangle, but with half its height. The proof has now been reduced to simply noting that these geometric figures have equal area. For instance, if we double the triangle by flipping it over its hypotenuse, and double the rectangle by flipping it over its upper border, we get the same figure.

Oresme also applied his graphical reasoning to invent a law usually attributed to Galileo: that the distance covered by an object under uniform acceleration grows with the square of the time. To see this, consider again the right triangle that is the area under the graph describing uniform acceleration. Its area is proportional to the product of its base and its height, both of which are proportional to time.

In his understanding of the nature of space, Oresme's instincts were equally astounding. Another scoop he got on Galileo was a component of Einstein's doctrine of relativity. It is the doctrine that only relative motion has meaning. Oresme's teacher in Paris, Jean Buridian, had argued that the earth could not rotate, for if it did, an arrow shot upward would come down in a different spot. Oresme rebutted this with his own example: a sailor at sea drawing his hand down along the mast perceives this motion as vertical. Yet to those of us on land, because the ship is moving, the motion of the man's hand will appear to move diagonally. Who is right?

Oresme asserted that the question itself is ill-defined: you cannot detect whether one body is moving except in reference to another. Today this is sometimes called *Galilean relativity.*

Oresme did not publish many of his works, nor did he carry them through to their logical conclusion. In many arenas, he approached the precipice of revolution, and then, for the sake of the church, he stepped back. For instance, based on his analysis of relative motion, Oresme went on to consider whether it might be possible to develop a theory of astronomy in which the earth rotated and even moved around the sun, the revolutionary ideas later promulgated by Copernicus and Galileo. But Oresme not only failed to convince his contemporaries, he eventually rejected his idea himself. His conversion came not from reason, but from the Bible. Quoting Psalm 93:1, Oresme wrote: "for God hath established the world, which shall not be moved."

On other issues, too, Oresme achieved brilliant insight into the nature of the world, then backed off from the truth he had perceived. For instance, on the existence of demons, he took a rebelliously skeptical view, asserting that their existence cannot be proved from natural law. Yet, ever the good Christian, he maintained that they exist as an article of faith. Perhaps marveling at his own equivocations, Oresme wrote, in the tradition of Socrates, "I indeed know nothing except that I know nothing." Rewarded for his faithfulness to the establishment, Oresme, who grew up poor, became royal adviser, ambassador, and tutor to Charles V. With Charles's support, Oresme was elevated to bishop in 1377, five years before his death.

Although there is no evidence that Galileo used any of Oresme's work directly, he was his intellectual heir. But Oresme's revolution in mathematics never truly bloomed, and the world had to wait another 200 years until, with the church weakened, two other Frenchmen cautiously took up the cause, this time to change the world of mathematics forever.

11. A Soldier's Story

N MARCH 31, 1596, a sick French no-blewoman with a dry cough, perhaps in-dicative of tuberculosis, gave birth to her third child. It was a weak, sickly baby. A few days later, the mother died. The doc-tors predicted that the baby would soon follow. It must have been a horrid time for the baby's father, but he did not give up. For the next eight years, he kept the child at home, in bed much of the time, attended by a nurse, and under his own loving care. The child would live for fifty-three years before the weakness of his lungs would finally bring him down. Thus was saved for the world one of its greatest philosophers and the architect of the next revolution in mathematics, René Descartes.

When Descartes was eight (some say ten), his father sent him to La Flèche, a Jesuit school that was then new but would soon become famous. The rector at the school allowed young Descartes to remain in bed until late each morning, when he felt ready to join the others. Not a bad habit if you can main-tain it, and this Descartes did until the last months of his life. Descartes did well at school, but upon finishing eight years later, he already exhibited the skepticism for which his phi-losophy would become famous: he was convinced that every-thing he had learned at La Flèche was either useless or mistaken. Despite that realization, at his father's wish, he spent the next two years engaged in more pointless learning, this time leading to a law degree.

Descartes finally abandoned the study of letters and moved to Paris. There, he spent his nights prowling the social circuit. By day, he lay in bed studying mathematics (beginning, of course, in the afternoon). He loved it, and it also occasionally brought him profit, as he applied his mathematics at the gambling tables. After a short time, though, Paris bored him.

What did a young man of independent means in Descartes's day do in order to travel and find adventure? He joined the army. In Descartes's case the army of Prince Maurice of Nassau. It was a true volunteer army: Descartes was not paid for his service. Prince Maurice got what he paid for. Not only did Descartes never see battle, but the following year he joined the opposing forces of the duke of Bavaria. This might seem pretty odd—first not fighting for one side, then not fighting for the other. But at the time, France and Holland's war with the Spanish-Austrian monarchy was in a hiatus. Descartes had joined the army to travel, not for political reasons.

Descartes enjoyed his days in the army, meeting people of different lands, yet finding the solitude he craved in order to study mathematics and science and ponder the nature of the universe. His travels almost immediately bore fruit.

One day in 1618, soldier Descartes was in the small town of Breda in Holland when he saw a crowd gathered around a notice posted on the street. He wandered over and asked an elderly onlooker to translate for him into French. There are many things such a notice might be today—an advertisement, a No Parking sign, a wanted poster. One kind of notice you will not find on streets today is what this actually was: a mathematical challenge to the public.

Descartes considered the problem and remarked offhand that he thought it quite easy. His translator, perhaps irked, perhaps amused, called the stranger's bluff and challenged him to solve it. Descartes did. The older man, a fellow named Isaac Beekman, was impressed, not an easy feat, for it turns out this bystander was one of the greatest Dutch mathematicians of his time.

Beekman and Descartes became such good friends that

Descartes later described Beekman as "the inspiration and spiritual father of my studies." It was Beekman to whom, four months later, Descartes first described his revolutionary way of looking at geometry. Letters from Descartes to his friend over the next couple years are generously peppered with references to his new realization of the relationship between numbers and space.

Throughout his life, Descartes was very critical of the works of the Greeks in general, but their geometry especially irked him. It could get awkward and appear needlessly difficult. He seemed to resent the fact that, the way Greek geometry was formulated, he had to work harder than necessary. In his analysis of a problem posed by the ancient Greek Pappus, Descartes wrote that "it already wearies me to write so much about it." He criticized their system of proofs because each new proof seemed to provide a unique challenge, which could be overcome, as Descartes wrote, "only on the condition of greatly fatiguing the imagination." He also disapproved of the way the Greeks had defined curves, by description, which could indeed get tedious and make proofs rather cumbersome. Today, scholars write that "Descartes's mathematical laziness is notorious," but Descartes was not ashamed to be seeking to find an underlying system that would make proving geometric theorems less taxing. That's how he could sleep in every day and still have more impact than any of those more industrious scholars who have criticized him.

As an example of Descartes's success, compare Euclid's definition of a circle from Part I with Descartes's:

Euclid: A circle is a plane figure contained by one line [i.e., curve] such that all straight lines falling upon it from one point amongst those lying within the circle—called the center—are equal to one another.

81

Descartes: A circle is all x and y satisfying $x^2 + y^2 = r^2$ for some constant number r.

Even to those who don't know what the equation means, Descartes's definition has to look simpler. The point isn't the interpretation of the equation, but merely that in Descartes's method the circle is defined by one. Descartes translated space into numbers, and, more importantly, used his translation to phrase geometry in terms of algebra.

Descartes began his analysis by turning the plane into a kind of graph by drawing a horizontal line called the "x-axis" and a vertical line called the "y-axis." Except for one important detail, any point in the plane is then described by two numbers: its vertical distance from the horizontal axis, called y, and its horizontal distance from the vertical axis, called x. These points are usually written as the "ordered pair" (x,y).

Now, the detail: if we literally measured distance as described above, there would be more than one point for every pair of coordinates (x,y). For instance, consider two points that are each one unit above the x-axis, but which lie on opposite sides of the y-axis, say one point being two units to the right, and the other two units to the left. Since both points lie one unit above the x-axis and both sit two units from the y-axis, according to our description, both would be described by the coordinate pair $(2,1)$.

This same ambiguity is possible with street addresses. Might two people, each living at 137 Eightieth Street, both stick their nose in the air and remark, "I would never live in *that* neighborhood." Why not? *West Side Story* and *East Side Story* are indeed two very different stories. Mathematicians dispose of the coordinate ambiguity the same way urban planners fix street addresses, only they use plus and minus signs instead of east/west and north/south designations. Mathematicians attach a minus sign to the x-coordinates of all the points to the left of the y-axis (i.e., the "east side"), and to the y-coordinate of all points which lie below the x-axis (i.e., the "south side"). In our example, the first point would keep its designation (2,1), but the second would instead be written (–2,1). This is like dividing the plane into four quadrants—northeast, northwest, southeast, and southwest. All points in a "south" quadrant have a negative y value, and all points in a west quadrant have a negative x value. This system of labeling is today called *Cartesian coordinates.* (Actually, it was invented at about the same time by Pierre de Fermat, but while Descartes had the bad habit of not including citations in his publications, Fermat had the worse habit of not even publishing.)

Of course, as we have seen, the use of coordinates alone was not new. Ptolemy had used them in his maps in the second century. Yet Ptolemy's work was merely geographic. He saw in it no significance beyond the globe. The real advance in Descartes's idea of coordinates came not in the idea of coordinates themselves, but in the use Descartes made of them.

In studying the classical Greek curves whose mode of definition Descartes so despised, Descartes found surprising patterns. For instance, he plotted a number of lines and found that for any line he drew, the x and y coordinates of every point on the line were always related in the same simple way. Algebraically, the relation is expressed by an equation of the form $ax + by + c = 0$, where a, b, and c were constants, plain numbers like 3, or 4½, that depended only upon the particular line he examined. This means that any point described by the pair (x,y) is on the line if, and only if, the sum of a times x, b times y, and c, equals zero. It is an alternative, algebraic definition of a line.

In Descartes's view, a line is a set of points with the property that if you increment one coordinate, to obtain another point in the set you must increment the other coordinate in a fixed ratio. His definition of the circle (or ellipse) works on the same principle, only when you take away from one coordinate, you have to add to the other so that the (weighted) sum of the squares of the coordinates, not simply the sum of the coordinates themselves, remains the same.

Three hundred years earlier, Oresme had also noticed that curves could be defined by relations between coordinates, and also derived a form of the equation for a line. But in Oresme's day, algebra had not been widely disseminated, and in the absence of better notation, Oresme couldn't carry the idea much further. Descartes's method of associating algebra and geometry amounts to a generalization of Oresme's ideas so that all curves in Greek mathematics could now be described simply and concisely. Ellipses, hyperbolas, parabolas, all proved to be definable via simple equations among their x- and y-coordinates.

The fact that classes of curves can be defined by an equation has far-reaching consequences in science. For example, below are Nicolai's data again, with the decimal point moved over one place. This reveals what they truly are: a table of the approximate average high temperatures on the 15th of each month (except January) in New York City. A scientist might ask: Is there a simple relation among the data?

Date	Average high temperature
2/15	40
3/15	50
4/15	62
5/15	72
6/15	81
7/15	85
8/15	83
9/15	78
10/15	66

11/15	56
12/15	40

As we have seen, when graphed, the data in this table form a simple geometric curve, the parabola. Knowledge of the equation that defines the parabola now gives us certain predictive powers—it allows us to formulate a "law of average highs" for New York City weather. The law is this: Let y stand for the number of degrees less than 85 degrees; let x stand for the number of months away from July 15; then y equals twice the square of x.

Let's try the law. To find the average high temperature in New York on, say, October 15, you note that October is three months after July, so x is 3. Since the square of 3 is 9, the average temperature on October 15 is twice 9, or 18, degrees less than the July 15 average high of 85. Thus, according to the "law," the average high is approximately 67 degrees. The actual average high is 66. For most months, the law works pretty well, and can also be used for days other than the 15th of the month if you don't mind working with fractions.

The law of average highs defines a relation between y and x; this is a special case of what mathematicians call a function. In this case, the parabola is the graph of the function. Physical science is largely concerned with doing what we just did: noticing regularities in data, discovering the functional relationships, and (which we did not do) explaining the cause for them.

Just as physical laws can be inferred graphically employing Cartesian methods, Euclid's theorems, too, have algebraic consequences. For instance, think about the Pythagorean theorem in Cartesian terms. Imagine a right triangle. For simplicity, let's imagine that it has a vertical side along the y-axis stretching from the origin up to a point A, and a horizontal side along the x-axis stretching from the origin

85

over to a point B. Then the length of the vertical side is simply the y-coordinate of its endpoint, A, and the length of the horizontal side is the x-coordinate of its endpoint, B.

The Pythagorean theorem in this case tells us that the sum of the squares of the horizontal side and the vertical side, $x^2 + y^2$, is the square of the length of the hypotenuse. If we accept the definition that the distance between two points such as A and B is the length of the line connecting them, then we have just found that the square of the distance between A and B is $x^2 + y^2$. But now consider any two points A and B in the plane. We can choose to draw our x- and y-axes so that we have the situation just described—with A along the horizontal axis and B along the vertical axis. That means that the square of the distance between any two points, A and B, is simply the sum of the squares of their horizontal and vertical separations.

● ■ ▲

Descartes's formula for distance is deeply tied to Euclidean geometry, as we'll see later. But his way of looking at distance as a function of coordinate differences is a generally valid concept, one that later became a key to understanding the nature of both Euclidean and non-Euclidean geometries.

Descartes utilized his geometric insight to do famous work in many areas of physics. He was the first to formulate the law of the refraction of light in its present trigonometric form; he was also the first to explain completely the physics of the rainbow. His geometric methods were so crucial to his insights that he wrote that "my entire physics is nothing other than geometry." Yet Descartes delayed publishing coordinate geometry for nineteen years. In fact, he didn't publish anything until he was forty. What was he afraid of? The usual suspect, the Catholic Church.

After the repeated urging of friends, Descartes had been on the brink of publishing a book a few years earlier, in 1633. Then this Italian fellow named Galileo published a book called *Dialogue on the Two Chief Systems*. It was a cute piece involving three talking heads in a dialogue about astronomy. Definitely Off-Broadway. But for some reason the church fathers decided to review it, and they weren't very impressed. Perhaps they thought the actor representing their Ptolemaic point of view didn't get enough good lines. Unfortunately, in those days, when the church reviewed a book, it also reviewed the author, who was as likely as the book to be the subject of a subsequent bonfire. In Galileo's case, only the book was burned. Galileo himself was forced to renounce it, and, oh yeah, he also got an open-ended term of imprisonment imposed by the Inquisition. Descartes was not a fan of Galileo's. In fact, in a letter, he wrote his own review of Galileo's work: "It seems to me that he [Galileo] lacks a great deal in that he is continually digressing and never stops to explain one topic completely, which demonstrates that he has not examined them in an orderly fashion. . . ." Yet he shared Galileo's heliocentric point of view, and other rational ideas, and he took Galileo's condemnation to heart. Even though he lived in a Protestant country, he canceled publication.

Descartes finally regained his courage and published his first work in 1637, taking care to make his book as inoffensive as possible to the church. By the age of forty, Descartes had much more than just geometry to communicate, and he tacked much of it together in this single book. The preface alone had 78 pages. The original manuscript had the not-too-snappy title: *Project for a Universal Science Which Might Raise Our Nature to Its Highest Degree of Perfection. Next, the Dioptric, the Meteors, and the Geometry, Where the Most Curious Matters Which the Author Could Find to Give Proof of the Universal Science He Proposes Are Explained in Such a Manner That Even Those Who Have Never Studied Can Understand Them.* At publication, the title was shortened a little, presumably by the seventeenth-century equivalent of the marketing department. Still, it was rather lengthy. Time has further eroded its length, and today the work is generally referred to as *Discourse,* or *Discourse on Method.*

Discourse on Method was a long essay describing Descartes's philosophy and his rational approach to solving problems in science. *Geometry,* the third appendix, was intended to show the results his approach could achieve. He kept his name off the title page, not because the title left no space, but because he still feared persecution. Unfortunately, his friend Marin Mersenne wrote an introduction that left no doubt as to the identity of the book's author.

As he had feared, Descartes was roundly attacked for his perceived challenge to the church. Even his mathematics attracted nasty criticism. Fermat, who as we said had discovered a similar algebraization of geometry, objected to trivial points. Blaise Pascal, another brilliant French mathematician, condemned it completely. Yet personal feuds can hold back scientific advances only temporarily, and within a few years Descartes's geometry was part of the curriculum in almost every university. His philosophy was not as readily accepted.

Descartes was most viciously attacked by a man named Voetius, the head of the divinity department at the University of Utrecht. Descartes's heresy, according to Voetius, was the usual one, his belief that reason and observation could determine the truth. Descartes in fact went even further, believing that people could control nature, and that cures for all disease and even the secret of eternal life would soon be found.

Descartes had few friends, and never married. He did, however, have one affair in his life, with a woman named Helen. With her, he had a child, Francine, in 1635. It is believed that the three of them lived together between 1637 and 1640. In the fall of 1640, in the midst of his battle with Voetius, Descartes went away to tend to the publication of a new work. Francine became ill, with purple blotches erupting all over her body. Descartes rushed back home. We don't know if he arrived in time, but she died on the third day of her illness. Descartes and Helen soon ended their relationship. If not for a record of her life and death written into the flyleaf of one of Descartes's books, we would never have known that Francine was his daughter, and not, as he said in order to avoid scandal, his niece. Though throughout his life Descartes was famous for his lack of emotion, this loss devastated him. He would barely survive the decade.

12. Iced by the Snow Queen

FEW YEARS after Francine's death, twenty-three-year-old Queen Christina of Sweden invited Descartes to her court. Christina was played by Greta Garbo in her 1933 film biography, and the idea of an elegant young Swedish woman perhaps does evoke the image of a tall blithe blonde. As usual, the Hollywood history wasn't quite true to the facts. The real Christina was short, with uneven shoulders and a deep masculine voice. She disliked customary women's clothing, and was described by some as resembling a cavalry officer. It was said that, as an infant, she enjoyed the sound of gunfire.

By the age of twenty-three, Christina was already a harsh taskmaster, with little tolerance for wimps. She slept but five hours a day, and didn't shiver at the thought of long stretches of frigid Swedish winter when you could have played ice hockey on hosed-down asphalt (had hoses, hockey, or asphalt been invented). Even hundreds of years later, those of us reading about Descartes can guess that her court was probably not his cup of iced tea. Yet Descartes did go. Why?

Christina was a brilliant woman, dedicated to learning, who felt isolated in her northern country. With the aim of creating in her snowy land an intellectual paradise, a center of learning far from the center of Europe, she spent large sums of money acquiring volumes for a grand library. If, like Ptolemy, she collected books, unlike him, she also collected their authors. Descartes's fate was sealed when he met and befriended Pierre Chanut in 1644. The next year, Chanut was sent to Sweden as a minister of the king of France. In Sweden, he tooted his friend's horn, and to his friend, he sang the praises of the Snow Queen. Christina agreed with Chanut that

Descartes would be a premier catch. She sent an admiral to France to persuade Descartes to come. She promised Descartes that which was closest to his heart: to build for him an academy, of which he would be the director, and a house in the warmest part of Sweden (which, in retrospect, wasn't promising much). Descartes wavered, but finally accepted. He had no access to weather.com, but he certainly knew of the climate—and the personality—that awaited him. The day before he left, he wrote his will.

The winter that met Descartes in 1649 was one of the harshest in the history of Sweden. If he had entertained the idea of lying all day under multiple thick blankets, warm and cozy and protected from the freezing cold as he pondered the nature of the universe, Descartes was soon given a rude wake-up call. He was summoned to appear each day at 5 a.m. in Christina's court to give Christina her daily five-hour lesson in morality and ethics. Descartes wrote to a friend: "It seems to me that men's thoughts freeze here in winter just like the water. . . ."

That January, his friend Chanut, with whom he had been staying, became ill with pneumonia. Descartes helped nurse his friend to recovery, but in the process came down with the disease himself. Descartes's doctor was away, so Christina sent another physician, who happened to be an avowed enemy of Descartes, who had made many of the Swedish court angry with jealousy. Descartes refused to be treated by the man, who in any case probably wouldn't have helped—his prescribed treatment was to bleed Descartes. Descartes's fever steadily rose. Over the next week or so, he suffered from bouts of delirium. In between, he spoke of death and philosophy. He dictated a letter to his brothers asking them to look after the nurse who had taken care of him in his fragile childhood. A few hours later, on February 11, 1650, Descartes died.

He was buried in Sweden. In 1663, the goal of Voetius's attacks was finally realized: the church banned Descartes's writings. But the church had weakened by then to the degree that in many circles this only added to his popularity. The government of France requested that Descartes's remains be returned, and in 1666, after much pleading, the Swedish government shipped back his bones. Well, most of them: they kept his skull. Descartes's remains were moved several more times. Today they are marked by a small commemorative stone in Saint-Germain-des-Près. That is, except for his skull, which was finally returned to France in 1822. Today it can be viewed in a glass case in the Musée de l'Homme.

Four years after Descartes's death, Christina abdicated her throne. She converted to Catholicism, crediting Descartes and Chanut for her enlightenment. She eventually settled in Rome, perhaps learning from Descartes also the advantages of warmer climes.

III

THE STORY

OF GAUSS

Can parallel lines intersect in space? Napoleon's favorite prodigy presents Euclid with his Waterloo. The greatest revolution in geometry since the Greeks begins.

13. The Curved Space Revolution

E UCLID aimed to create a consistent mathematical structure based on the geometry of space. The properties of space derived from his geometry are therefore the properties of space as the Greeks understood it. But does space really have the structure described by Euclid and quantified by Descartes? Or are there other possibilities?

We don't know if Euclid would have raised an eyebrow had he been told that his *Elements* would remain sacrosanct for 2,000 years, but as they say in the software business, 2,000 years is a long time to wait for version 2. A lot changed in that time: we discovered the structure of the solar system; we gained the ability to sail around, and map, the globe; we stopped drinking diluted wine for breakfast. And, in that time, the mathematicians of the Western world had developed a universal aversion to Euclid's fifth postulate, the parallel postulate. Alas, it was not the content they found distasteful, it was its place as an assumption rather than a theorem.

Through the centuries, the mathematicians who attempted to prove the parallel postulate as a theorem came close to discovering strange and exciting new kinds of space, but each of them was hampered by a simple belief: that the postulate was a true and necessary property of space.

All but one, that is, a teen-aged boy of fifteen named Carl Friedrich Gauss, who, as it happened, would become one of Napoleon's heroes. With this young genius's realization in 1792, the seeds of a new revolution were planted. Unlike the previous ones, this would not be a revolutionary improvement on Euclid, it would be an entirely new operating system.

Soon the strange and exciting spaces overlooked for so many centuries were discovered and described.

With the discovery of curved spaces came the natural question, is our space Euclid's, or one of those others? That question eventually revolutionized physics. Mathematics, too, was thrown into a quandary. If Euclid's structure isn't simply an abstraction of the true structure of space, then what is it? And if the parallel postulate can be questioned, what about the rest of Euclid's edifice? Soon after the discovery of curved space all of Euclidean geometry came tumbling down, and then— surprise! The rest of mathematics tumbled as well. By the time the dust cleared, not only the theory of space, but physics and mathematics, too, were in a new era.

To understand how difficult a leap it was to contradict Euclid, one has to appreciate how deeply entrenched was his description of space. Already in his own, ancient time, Euclid's *Elements* was a classic. Euclid not only defined the nature of mathematics, but his book played a central role as a model of logical thought in education and natural philosophy. It was a key work in the intellectual revival of the Middle Ages. It was one of the first books printed after the invention of the printing press in 1454, and from 1533 until the eighteenth century it was the only one of all the Greek works to exist as a printed text in the original language. Until the nineteenth century, every work of architecture, the composition in every drawing and every painting, every theory and every equation employed in science were all inherently Euclidean. *Elements* was not unworthy of its great stature. Euclid transformed our intuition of space into an abstract logical theory from which we could make deductions. Perhaps most of all, we must credit Euclid with attempting to shamelessly bare his assumptions, and never pretending that the theorems he proved were anything more than logical deductions from his few unproven postulates. As we saw in Part I, though, one of these postulates, the parallel postulate, caused consternation in almost every scholar who studied Euclid because it was not as simple and intuitive as Euclid's other assumptions. Recall its wording:

> Given a line segment that crosses two lines in a way that the sum of inner angles on the same side is less than a right angle, then the two lines will eventually meet (on that side of the line segment).

Euclid didn't use the parallel postulate at all in proving his first twenty-eight theorems. By then he had already proven the converse of the postulate, as well as other statements that seemed far better candidates for "axiomhood"—like the fundamental fact that the lengths of any two sides of a triangle have to add up to more than the length of the third. Why, then, so far down the road, did he need to introduce such an arcane, technical postulate? Did he write that chapter on deadline?

For over 2,000 years, as 100 generations lived and died, as borders changed and political systems rose and fell, as the earth hurtled 1,000 billion miles around and around our sun, thinkers everywhere remained dedicated to Euclid, questioning their god not on any issue of content, but only on this one teeny point: couldn't the ugly parallel postulate be proved?

14. The Trouble with Ptolemy

HE FIRST known attempt to prove the parallel postulate was made by Ptolemy in the second century A.D. His reasoning was complicated, but in essence his method was simple: he assumed an alternate form of the postulate, then derived the original form from it. What should we think of Ptolemy? Did he live in an intelligence-free zone? Should we picture him racing to his friends, exclaiming, "Eureka! I found a new form of proof: the circular argument." Mathematicians would not make the same mistake twice. They would make it over and over. For as it turns out, some of the most innocuous assumptions, some so obvious as to be left unstated, have in the end proven to be the parallel postulate in disguise. The connection of the postulate to the rest of Euclidean theory is as subtle as it is deep. A couple hundred years after Ptolemy, Proclus Diadochus made the next notable attempt to prove the postulate once and for all. Proclus was educated in Alexandria in the fifth century, and then moved to Athens, where he became head of Plato's Academy. Proclus spent long hours analyzing Euclid's work. He had access to books that have long since disappeared from the face of the earth, such as *History of Geometry* by Eudemus, a contemporary of Euclid's. Proclus wrote a commentary on book 1 of *Elements* that is the source for much of our knowledge of ancient Greek geometry.

To understand Proclus' argument, it is useful to do three things: First, use an alternate form of the parallel postulate given earlier, Playfair's axiom. Second, make Proclus' argument a little less technical. And third, translate from the Greek. Playfair's axiom is this:

Given a line and an external point (a point not on the line), there is exactly one other line that passes through the external point and is parallel to the given line.

In today's world, most of us find maps and streets a lot more understandable than lines labeled with obscure symbols like α or λ. So to put Proclus' argument in a more relevant setting, imagine, say, Fifth Avenue in New York City. Next, imagine another avenue, parallel to Fifth, which we will call Sixth Avenue. Remember that by parallel, we mean, according to Euclid, "does not intersect," so our assumption is that Fifth Avenue does not intersect Sixth Avenue.

Rising high above the coffee vendors and hot dog stands on Sixth Avenue is a venerable building housing that esteemed publisher of only the highest quality books, The Free Press (coincidentally, also the publisher of this book). Not to diminish them, but in this example, The Free Press will play the role of the "external point."

Now, in mathematical tradition, keep in mind that what we have just stipulated is *all* we can assume about these roads. Although for purposes of concrete illustration we have specific avenues in mind, as mathematicians, we may use no properties of those avenues in this proof other than those we explicitly state. So if you happen to know that an also-ran publisher (at least regarding this book) named Random House is just down the street, that Fifth Avenue and Sixth Avenue are a certain distance apart, or that a specific corner is inhabited by a drooling psychopath, put those thoughts out of your mind. A mathematical proof is an exercise in employing only the explicitly granted facts, and none of the properties of New York City is mentioned in Euclid's *Elements*. It is, in fact, an unjustified assumption you will probably make without thinking that makes the following argument of Proclus false.

We are ready to state Playfair's axiom in the form that applies to our setup:

> Given Fifth Avenue and a publisher named The Free Press on Sixth Avenue, there can be no other roads that run by The Free Press that, like Sixth Avenue, are parallel to Fifth Avenue.

This statement isn't exactly equivalent to the Playfair axiom, because, like Proclus, we have assumed that at least one line, or road (Sixth Avenue), parallel to the given one (Fifth Avenue), exists. This actually must be proved, but Proclus interpreted one of Euclid's theorems as guaranteeing it. We will accept that for now and see if, following his argument, we can prove the axiom in the above form.

To prove the postulate, i.e., to make it a theorem, we must show that any road passing The Free Press other than Sixth Avenue must intersect Fifth Avenue. This seems obvious from our daily experience—which is why such a road is called a cross street. All we have to do here is prove it without utilizing the parallel postulate. We begin by imagining a third road, whose only assumed properties are that it is straight and runs past The Free Press. Let's call that road Broadway.

In his method of proof, Proclus would start at The Free Press and walk downtown along Broadway. Imagine a street running from where Proclus happens to stand to Sixth Avenue, perpendicular to Sixth Avenue. Call this new street Nicolai Street. The setup is illustrated on the following page.

Nicolai Street, Broadway, and Sixth Avenue form a right triangle. As Proclus walks farther downtown along Broadway, the right triangle formed in this way gets larger and larger. Eventually, the sides of the triangle, including Nicolai Street, get as long as you like. In particular, the length of

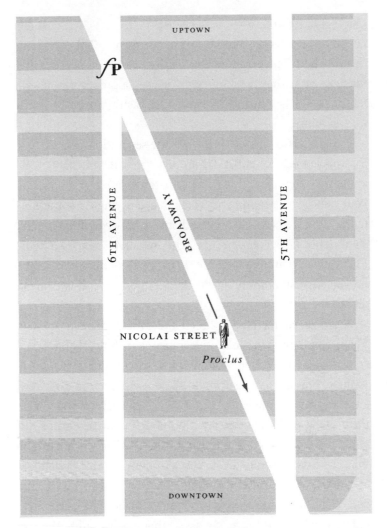

PROCLUS'S PROOF

Nicolai Street eventually exceeds the separation of Fifth and Sixth avenues. Hence, Proclus would say, Broadway must cross Fifth Avenue, which is what was to be proved.

This argument is simple, but false. For one thing, there is a subtle misuse of the idea "larger and larger." Nicolai Street could become ever longer without ever becoming longer than one block, like the series of numbers ½, ⅔, ¾, ⅘, ⅚ . . . which becomes ever larger, but never greater than 1. This flaw can be remedied. The essential flaw is that, like Ptolemy, Proclus made an unjustified assumption. He used a property of parallel roads that is intuitive but which he hasn't proved. What is he assuming?

Proclus' error was in his use of "the separation of Fifth and Sixth Avenue." Recall the warning, ". . . if you happen to know that they are . . . a certain distance apart, put it out of your mind." Although Proclus is not specifying what the distance is, he is asserting that the distance between them is constant. This is our experience with parallel lines, and with Fifth and Sixth Avenue, but it cannot be mathematically proven without employing the parallel postulate: it is equivalent to the postulate itself.

A similar point also stymied the great Baghdad scholar Thābit ibn Qurrah in the ninth century. To follow Thābit's reasoning, imagine Thābit walking in a straight line along Fifth Avenue, holding a measuring stick perpendicular to Fifth Avenue and one N.Y. avenue–block long. As Thābit walks down Fifth, what path is traced by a point at the other end of his stick? Thābit would have asserted that this path is a straight line, say Sixth Avenue. From this assumption, Thābit went on to "prove" the parallel postulate. The line traced by the far end of the measuring stick is certainly a curve of some sort, but by what authority can we assert that it is a straight line? It turns out that that authority can only be—you guessed it—the parallel postulate. Only in Euclidean space is the set of points equidistant from a line also a line. Thābit, too, had repeated Ptolemy's mistake.

Thābit's analysis touches on deep issues in the concept of space. Euclid's system of geometry depends on being able to move figures around and superimpose them. That is how you check the congruence, or equivalence, of geometric shapes.

Imagine that you want to move a triangle. The natural way to accomplish this is to take its three sides, each of which is a segment of a line, and displace them by identical amounts in an identical direction. But if the set of points equidistant from a line is not also a line, this means that the sides of the displaced triangle will not be lines. As it moves, the figure will become distorted. Could space actually have this property? Unfortunately, instead of following this reasoning to the wonderful place it leads, Thābit interpreted the specter of distortion as "proof" that his assumption about the equidistance of lines must be justified.

Not long after Thābit, Islamic support for the sciences dwindled. In one locale a scholar even complained that where he lived it was legal to kill mathematicians. (This was probably due less to a disdain for nerds than to the mathematicians' habit of studying astrology, which through history was often connected with black magic and considered dangerous rather than amusing as it is today.)

The year in the Christian calendar was to nearly double before the geometric work of Thabit and his followers was resurrected. That happened when, in 1663, the English mathematician John Wallis gave a lecture quoting one of Thābit's successors, Nāsir Eddin al-Tusi.

Wallis was born in Ashford, Kent, in 1616. When he was fifteen, he saw his brother reading a book on arithmetic and became fascinated with the subject. Though he went on to study divinity at Emmanuel College, Cambridge, and in 1640 was ordained as a priest, he remained devoted to mathematics. It was the time of what is usually called the English Civil War, a struggle with religious overtones between the king, Charles I, and the English Parliament. Wallis was skilled at cryptography, the branch of mathematics that deals with decoding messages, and employed his skills to aid the Parliamentarians. It is for this, some say, that he was granted the

Savilian Chair of geometry at Oxford in 1649 after his prede-
cessor, Peter Turner, was dismissed for his royalist views. For
whatever reason, for Oxford it was a good swap.

Turner had never been more than a crony of the archbishop
of Canterbury, squaring all the right political circles but never
once publishing a mathematics paper. Wallis became the leading English mathematician of the pre-Newtonian era, and an
important influence on Newton himself. Today, even non-
mathematicians, especially those owning a certain brand of
luxury automobile, are familiar with one aspect of his work:
he introduced the symbol ∞ for infinity.

Wallis's idea for reforming Euclidean geometry was to replace Euclid's distasteful parallel postulate with an intuitively
obvious one that could be put this way:

> Given any side of any triangle, the triangle can be blown
> up or shrunk so that the chosen side has any length you like,
> but the angles of the triangle are not altered.

For instance, if you have a triangle whose angles are each
60 degrees, and whose sides are each one unit long, then you
assume that there exists another triangle whose angles are also
60 degrees, but whose sides are anything you like: 10, 10, 10,
or 1/10, 1/10, 1/10, or 10,000, 10,000, 10,000. Such triangles, with
sides proportionally larger or smaller, but corresponding angles equal, are called similar triangles. If we assume Wallis's
axiom, then, ignoring a couple of technical details that can be
overcome, the parallel postulate is now easy to prove employing reasoning similar to that of Proclus. Wallis's "proof" never
gained acceptance among mathematicians because all it really
did was to substitute one postulate for another. But, reversing
Wallis's reasoning leads to an amazing statement: If a space
exists in which the parallel postulate does not hold, then *no*
similar triangles exist.

Who cares? Well, the trouble is, triangles are everywhere. Cut a rectangle along its diagonal and you have two triangles. Put your hand on your hip and your bent arm forms a triangle with the side of your body. In fact, though everyone's body is different, your body, and most objects, can be modeled to a good approximation as a lattice of triangles: this is the principle behind 3-D computer graphics. If similar triangles do not exist, then many of the assumptions of our everyday life will not hold true. Look at a cute pantsuit in a clothing catalogue and you assume that what arrives in the mail will match the image, though it may be dozens of times as large. Fly on your favorite airline and you are exhibiting faith that the wing shape that seemed to work in scale models will have the same nice properties as part of a huge jet. Hire an architect to add a few rooms to your house, and you expect the addition to match the blueprint. In non-Euclidean space, none of this would be true. Your clothes, the airplane, your new bedroom would come out distorted.

Perhaps such bizarre spaces exist mathematically, but can real space have these properties? Wouldn't we have noticed? Maybe not. A deviation of 10 percent in the shape of your smile might catch your mother's notice, but not a difference of 0.0000000001 percent. Non-Euclidean spaces are almost Euclidean for figures that are small—and we live in a relatively *small* corner of the universe. Like quantum theory, in which the laws of physics take on bizarre new forms, but only in realms far smaller than we encounter in daily life, curved space might exist, yet be so nearly Euclidean that on the scale of normal terrestrial life we don't notice the difference. And yet, like quantum theory, the implications of curvature for the theories of physics could be enormous.

By the end of the eighteenth century, if mathematicians had viewed their discoveries differently, they would have concluded that non-Euclidean spaces might exist, and that if they did, they would have some very strange properties. Instead, mathematicians were merely frustrated that they couldn't prove that these strange properties lead to a contradiction, and therefore that space is Euclidean.

The next fifty years were years of a secret revolution. Gradually, in several countries, new kinds of space were discovered, but they were either not revealed or not noticed by the community of mathematicians. Not until scholars studied the papers of a recently deceased old man in Göttingen, Germany, in the middle of the nineteenth century did the secrets on non-Euclidean space become known. By then, most of those who had discovered them, like him, were dead.

15. A Napoleonic Hero

N GÖTTINGEN, on February 23, 1855, the man who was at the center of the assault on Euclid lay in his cold bed, old, and fighting for every breath. His weak heart could barely pump his blood, and his lungs were filling with fluid. His pocketwatch tick-tick-ticked away his remaining time on earth. It stopped. At almost the same moment, so did his heart. It was the kind of symbolic touch that normally only novelists get to employ.

A few days later, the old man was buried next to the unmarked grave of his mother. After his death, a sizable fortune in money was found hidden all over his house—stuffed into desk drawers, cabinets, desks. His house was modest, his tiny study furnished only with a small table, a desk and a sofa, and lit by a single light. His small bedroom had no heat.

For most of his life he had been an unhappy man, with few close friends and a deeply pessimistic outlook on life. He had spent decades teaching in the university, yet he considered it "a burdensome, ungratifying business." He felt that "without immortality, the world would be meaningless," yet he could not convince himself to become a believer. He had won many honors, yet of his honors he wrote that "the griefs overbalance the joys a hundredfold." He was at the fulcrum of the revolution against Euclid, yet he never wanted it revealed. To scholars of mathematics, then and now, this man is considered, with Archimedes and Newton, to have been one of the greatest mathematicians in the history of the world.

Carl Friedrich Gauss was born in Braunschweig (Brunswick), Germany, on April 30, 1777, fifty years after Newton died. He came from a poor neighborhood in a squalid city, about 150 years past its prime. His parents belonged to a class of populace called, with German precision, "half-citizens." His mother, Dorothea, was illiterate and worked as a maid. His father, Gebhard, worked at various poorly paid servile jobs ranging from digging canals and bricklaying to keeping the accounts for a local funeral society.

A warning: sometimes when someone says that a person is "hardworking and honest," it is not a good sign. You get the feeling you are waiting for the other shoe to drop. *He was hardworking and honest. If only he hadn't kept his son bound and gagged in the closet those fourteen years* . . . Having forewarned the reader, one can safely say this: Gebhard Gauss was a hardworking and honest man.

There are many stories about Carl Gauss's childhood. He could do arithmetic almost before he could talk. One conjures up images of a toddler pointing to a street vendor's food stand, imploring his mother, "I hungry! I want!", then after the purchase crying because he can't figure out how to say, "He overcharged you by thirty-five cents." Apparently, this isn't far from the truth. The most famous story of Gauss's early talent took place on a Saturday around the time Gauss turned three. His father was adding up the weekly payroll for a group of laborers. The computation was taking a while, and Gebhard was unaware that his son was watching. Suppose Gebhard had had a normal two- or three-year-old named, say, Nicolai. What would typically happen at this point would be for Nicolai to spill a glass of milk all over the calculations and yell, "Sorry" and "I want more milk" in roughly the same breath. Instead, Carl said something like, "You added wrong. It should be . . ."

Neither Gebhard nor Dorothea had drilled the toddler in addition; in fact, nobody had taught Carl anything about arithmetic at all. To most of us, this behavior would thus seem about as natural as finding Nicolai at 2 a.m. sitting upright in bed speaking in ancient Aztec, as if he were possessed, if not by Satan, then at least by a kid over ten. But Carl's parents were used to it. By then, toddler Carl had already taught himself to read.

Unfortunately, Gebhard's idea of nurturing his son's talents wasn't to hire Carl a private tutor and send him to a Montessori school. This is understandable given that the family was poor and Maria Montessori wouldn't be born for another hundred years. Still, Gebhard might have found some way to encourage his son's education. Instead, he assigned Carl the weekly task of checking his payroll arithmetic, and occasionally pulled the toddler out to amuse his friends, a kind of one-kid freak show. Young Carl had poor eyesight, and sometimes couldn't read the slate of numbers his father set forth for him to add. Too shy to say anything, Carl just sat there and accepted failure. Before long, Gebhard sent Carl to work in the afternoons, spinning flax to supplement the family income.

In his later years, Gauss was openly scornful of his father, calling him "domineering, uncouth, and unrefined." Fortunately, Carl was blessed with two others in his family who did appreciate his gift: his mother and his uncle Johann, Dorothea's brother. Where Gebhard discounted his son's gifts and considered formal education pointless, Dorothea and Johann believed in his talent and fought Gebhard's resistance at every step. Carl was Dorothea's pride and joy from the moment he was born. Years later, Carl brought to his humble home a college buddy, Wolfgang Bolyai, who, though far from wealthy, happened to be a Hungarian nobleman. Dorothea took Carl's friend aside, and in a fashion that still seems thoroughly modern, asked if Carl was really as smart as everyone said, and if so, where it would lead him. Bolyai answered that Carl was destined to become the greatest mathematician in Europe. Dorothea burst into tears.

109

Carl entered his first school at the age of seven, his local grammar school. It wasn't anything like La Flèche, the Jesuit school Descartes entered at age eight that would later become famous. Instead, descriptions of Gauss's first school range from "squalid prison" to "hellhole." The squalid prison/hellhole/school was run by a warden/devil/school master named Buettner, whose name is apparently German for "Do as I say or I'll whip you." In his third year at the school, Carl finally was allowed to study the arithmetic he had already been capable of at age two.

In arithmetic class, Buettner enjoyed stimulating his young students' interest in mathematics by giving them tall columns of numbers to add, some up to 100 numbers long. Buettner apparently felt he himself was not worthy of doing such entertaining tasks, so he always assigned numbers he could easily sum employing one formula or another, formulae he kindly did not share with his class.

One day Buettner assigned the problem of adding all the numbers from 1 to 100. As soon as Buettner had finished stating the problem, his youngest pupil, Carl, turned in his slate. It was an hour before the others finished. When Buettner finally scrutinized the slates, he found that Carl was the only one in the class of fifty to add the numbers correctly, and Carl's slate didn't show a sign of any calculation. He had apparently figured out the summation formula and calculated the answer in his head.

It is speculated that Gauss discovered this by noticing what happens if you consider adding not one, but *two* sets of all the integers from 1 to 100. Then you can rearrange the addition in this way: add 100 and 1, 99 and 2, 98 and 3, and so on. You end up with 100 terms, each of which is equal to 101, so the sum of all the integers from 1 to 100 must be half of 100 times 101, or 5050. This is a special case of a formula that was known already to the Pythagoreans. In fact, they used it as a password in their secret society: the sum of the numbers from one to any number is equal to one half of the last number times the last number plus one.

Buettner was astounded. Quick as he was to use the whip on the laggards, he also appreciated genius. Gauss, who eventually taught mathematics in college, never himself whipped a student, but Buettner's attitude toward genius and his scorn for the lack thereof seems to be the one thing Buettner passed on to him. Years later, Carl would write in disgust about the three students in one of his classes, "one is only moderately prepared, the other less than moderately and the third lacks both preparation and ability. . . ." His comments about those three represent his general attitude about teaching. For their part, most of his students had equal scorn for his ability as a teacher.

With his own money, Buettner obtained from Hamburg the most advanced arithmetic textbook available. Perhaps Carl had finally found the mentor he desperately needed. Carl quickly read through the book. Unfortunately, it failed to challenge him. At this point, Buettner, as skilled an orator as he was a mathematician, proclaimed: "I can teach him nothing more," and gave up, presumably so he could focus once again on whipping his less gifted students who were beginning to feel neglected. Nine-year-old Carl was one step closer to a career of wurst breaks and heavy calluses.

But Buettner didn't leave Carl's genius completely unattended. He assigned his gifted seventeen-year-old assistant, Johann Bartels, to see what he could do. At the time, Johann had the fascinating job of making quill pens and teaching Buettner's students how to use them. Buettner knew that Bartels also had a passion for mathematics. Soon the nine- and seventeen-year-old were studying together, improving textbook proofs, helping each other discover new concepts. A few years passed. Gauss became a teenager. Anyone who has ever had a teenager, known a teenager, or been a teenager knows this can mean trouble. In Gauss's case, the only question was trouble for whom?

Today, being a rebellious teen might mean staying out all night with that girl with the diamond stud through her tongue. In Gauss's day, body piercing may have been left for the battlefield, but rebellion against mores was no less an "in" thing. The big intellectual movement in Germany at the time was called *Sturm und Drang,* or "storm and stress."

Any time a German social movement makes prominent use of the word *storm,* you have to watch out, but this one was led by figures such as Goethe and Schiller, rather than Hitler and Himmler. It preached worship of individual genius and rebellion against established rules. Though Gauss is not normally considered an adherent of the movement, he was a genius and he acted in accordance with it in his own way: he did not rebel against his parents or the political system; he rebelled against Euclid.

At twelve, Gauss began criticizing Euclid's *Elements.* He focused, as others had done, on the parallel postulate. But his criticism was new and heretical. Unlike all those before him, Gauss sought neither to find a more palatable form of the postulate nor to render it unnecessary by proving it from the others. Instead, he questioned whether it was valid. Is it possible, Gauss wondered, that space is actually curved?

By fifteen, Gauss became the first mathematician in history to accept the idea that a logically consistent geometry could exist in which Euclid's parallel postulate does not hold. This was still, of course, a long way from proving it, or creating such a geometry. Despite Gauss's talents, at fifteen he was still in danger of becoming just another ditch-digger. Fortunately for Gauss and science, his friend Bartels knew a guy who knew a guy who knew a guy named Ferdinand, duke of Brunswick.

Through Bartels, Ferdinand got word of a promising

young man with mathematical genius. The duke offered to pay his bills through college. This left Carl's father as his main stumbling block. Gebhard Gauss seemed to believe that the only way to get ahead was to keep digging those ditches. Here, Dorothea, who couldn't read any of the books her son wanted to study, took a stand. She stuck up for her son, and Carl was allowed to accept the offer. At fifteen, he entered the local Gymnasium, which is roughly equivalent to a high school. In 1795, at eighteen, he entered the University of Göttingen.

The duke and Gauss eventually became good friends. The duke continued to support him even after college. Gauss must have known this couldn't last forever. Rumor had it that the duke's generosity was draining his fortune faster than was good for him, and in any case the duke was already in his sixties and might not have as generous a successor. Still, the next dozen years were Gauss's most intellectually profound.

In 1804, he fell in love with a kind and cheerful young woman named Johanna Osthoff. Under her spell, Gauss, who would so often in his life seem arrogant and supremely self-assured, was humble and self-deprecating. He wrote of Johanna to his friend Bolyai:

> For three days, that angel, almost too celestial for this earth, has been my fiancée. I am superabundantly happy. . . . Her cardinal trait is a quiet devout soul without one drop of bitterness or sourness. Oh, she is much better than I. . . . I had never hoped for this bliss; I am not handsome, not gallant, I have nothing to offer except a candid heart full of devoted love; I despaired of ever finding love.

Carl and Johanna were married in 1805. The next year they had a baby boy, Joseph, and in 1808, a daughter, Minna. Their bliss didn't last.

In the fall of 1806, it wasn't illness but a wound from a musket ball in a battle against Napoleon that took the duke's life. Gauss stood at his window in Göttingen and could only watch as a wagon carrying his mortally wounded friend and benefactor drove past. Ironically, Napoleon would later spare the city itself from destruction because of the presence of Gauss, commenting that "the foremost mathematician of all time lives there."

The death of the duke naturally brought the Gauss family financial hardship. Those proved to be the least of their troubles. In the next few years Carl's father and his supportive uncle Johann both died. Then, in 1809, Johanna gave birth to their third child, Louis. Minna's birth had been difficult, but with Louis's birth Johanna and the child both became gravely ill. A month later, Johanna died. Not long after that, their newborn also died. In a short period Carl's life had been rocked with tragedy after tragedy. And it wasn't over: Minna, too, was destined to die at an early age.

Gauss soon remarried, and eventually had three more children. But for him, after Johanna's death, life never again seemed to bring much joy. To Bolyai, he would write: "It is true that in my life I have won much that the world honors. But believe me, my dear friend, tragedy has woven itself through my life like a red ribbon. . . ." Shortly before his own death in 1927, one of Carl's grandsons found among his grandfather's papers a letter, stained by traces of tears. On them, his grandfather had written:

> Lonesome, I sneak about the happy people who surround me here. If for a few moments they make me forget my sorrow, it comes back with double force. . . . Even the bright sky makes me sadder. . . .

16. The Fall of the Fifth Postulate

AUSS would not be considered one of the greatest mathematicians ever had he not had a deep influence on many fields of mathematics. Yet he is sometimes considered a transitional figure, capping the developments started by Newton rather than laying the groundwork for future generations. This is not true of his work on the geometry of space: it was the kind of work that would eventually keep mathematicians and physicists busy for a century. Only one thing stood in the way of his revolution. He kept his work secret.

When Gauss arrived as a student in Göttingen in 1795, he found a lively interest in the question of the parallel postulate. As a hobby one of Gauss's teachers, Abraham Kaestner, collected literature on the history of the postulate. Kaestner even had a student, Georg Kluegel, write as his dissertation an analysis of twenty-eight failed attempts to prove it. Yet neither Kaestner nor anyone else was open to what Gauss suspected: that the postulate might not hold. Kaestner even remarked that only a crazy person would doubt the postulate's validity. Gauss kept his thoughts to himself, though it turns out he noted his ideas in a scientific journal that wasn't discovered until forty-three years after his death. Later in life, Gauss would dismiss Kaestner, who dabbled in writing, as "the leading mathematician among poets and the leading poet among mathematicians."

Between 1813 and 1816, as a professor teaching mathematical astronomy at Göttingen, Gauss finally made the definitive breakthroughs that had been waiting since Euclid: he worked out equations relating the parts of a triangle in a new, non-Euclidean space whose structure we today call *hyperbolic geometry*. By 1824, Gauss had apparently worked out an entire theory. On November 6 of that year, Gauss wrote to F. A. Taurinus, a lawyer who dabbled quite intelligently in mathematics, "The assumption that the sum of the three angles [of a triangle] is less than 180° leads to a special geometry, quite different from ours [i.e., Euclidean], which is absolutely consistent, and which I have developed quite satisfactorily for myself. . . ." Gauss never published this, and insisted to Taurinus and others that they not make his discoveries public. Why? It wasn't the church Gauss feared, it was its remnant, the secular philosophers.

In Gauss's day, science and philosophy hadn't completely separated. Physics wasn't yet known as "physics" but "natural philosophy." Scientific reasoning was no longer punishable by death, yet ideas arising from faith or simply intuition were often considered equally valid. One fad of the day which particularly amused Gauss was called "table-rapping," in which a group of otherwise intelligent people would sit around a table with their hands placed in an arched position upon it. After a half hour or so, the table, as if bored with them, would start to move or turn. This was supposedly some sort of psychic message from the dead. Exactly what message the ghouls were sending is unclear, although the obvious conclusion is that dead people like to position their tables against the far wall. In one instance, the entire Heidelberg law faculty followed for some time as their table moved across the room. One pictures a bunch of bearded, black-suited jurists pacing alongside, struggling to keep their hands in their appointed spot, attributing the locomotion to occult animal magnetism rather than their push. This, to Gauss's world, was reasonable; the thought that Euclid had erred was not.

● ■ ▲

Gauss had seen too many scholars involved in time-consuming feuds with lesser minds to risk becoming embroiled in one himself. For instance, Wallis, whose work Gauss respected, had been involved in a bitter dispute with the English philosopher Thomas Hobbes over the best way to calculate the area of a circle. Hobbes and Wallis traded public insults for over twenty years, resulting in much valuable time spent writing pamphlets with titles like *The Marks of the Absurd Geometry, Rural Language, etc. of Doctor Wallis.*

The philosopher whose followers Gauss feared most was Immanual Kant, who had died in 1804. Physically, Kant was the Toulouse Lautrec of philosophers: stooped, barely five feet tall, with a badly deformed chest. He joined the University of Königsberg in 1740 as a theology student, but found he had a penchant for mathematics and physics. After graduating, he began to publish works of philosophy, and became a private tutor and sought-after lecturer. Around 1770, he began work on what would become his most famous book, *Critique of Pure Reason,* published in 1781. Kant, noting that geometers of the day appealed to common sense and graphical figures in their "proofs," believed that the pretense of rigor ought to be dispensed with, and intiuition embraced. Gauss held the opposite view—that rigor was necessary, and most mathematicians were incompetent.

In *Critique of Pure Reason,* Kant calls Euclidean space "an inevitable necessity of thought." Gauss did not dismiss Kant's work out of hand. He read it first, then dismissed it. In fact, Gauss is said to have read *Critique of Pure Reason* five times attempting to understand it, a lot of effort for a fellow who picked up Russian and Greek with less effort than it would take most of us to find the Χωριάτικη Σαλάτα on an Athens menu. Gauss's struggle becomes more understandable when you consider the clarity of writing that led to passages of

Kant's such as this one, on the distinction between analytic and synthetic judgments:

> In all judgments in which the relation of a subject to the predicate is thought (I take into consideration affirmative judgments only, the subsequent application to negative judgments being easily made), this relation is possible in two different ways. Either the predicate to the subject A, as something which is (covertly) contained in this concept A; or outside the concept A, although it does indeed stand in connection with it. In the one case I entitle the judgment analytic, in the other synthetic.

Today, mathematicians and physicists worry little about what a philosopher would think of their theories. Famous American physicist Richard Feynman, when asked what he thought of the field of philosophy, gave a concise answer consisting of two letters, a "b," and another letter that is usually employed to form the plural. But Gauss took Kant's work seriously. He wrote that the above distinction between analytic and synthetic theories "is such a one that either peters out in triviality or is false." Yet he would divulge these thoughts, like his theories on non-Euclidean space, only to those he trusted. In a quirk of history that has raised many eyebrows, although Gauss did not publish his breakthroughs of 1815–24, at about the same time two other men, both with connections to him, did.

● ■ ▲

On November 23, 1823, Johann (János) Bolyai, son of Gauss's longtime friend Wolfgang Bolyai, wrote to his father that he had "created a new, different world out of nothing," meaning that he discovered a non-Euclidean space. In the same year, in Kazan, Russia, Nikolay Ivanovich Lobachevsky explored the consequences of the violation of the parallel postulate in an unpublished textbook on geometry. Lobachevsky had been tutored by Johann Bartels, then a professor in Kazan. Both Wolfgang Bolyai and Bartels had had a long interest in non-Euclidean space, and had engaged Gauss in a discussion of his ideas.

Was it coincidence? The genius Gauss discovers a great theory and is happy to discuss it with friends, but refuses to publish. Shortly thereafter, friends and relatives of his friends of his come out of the woodwork claiming they have made the same great discovery. It was enough to inspire at least one song about Lobachevsky, with incriminating lyrics like "Plagiarize, let no one else's work evade your eyes . . ." But most historians today believe that it was more the spirit than the specifics of Gauss's work that was passed on, and that Bolyai and Lobachevsky did not know of each other's efforts, at least at the time.

Unfortunately, neither did anyone else. The quintessential obscure mathematicians, when they spoke, no one listened. It didn't help that when Lobachevsky published his work, it was in an unknown Russian-language journal called the *Kazan Messenger.* Or that Bolyai's was buried in an appendix to one of his father's books, *Tentamen.* Some fourteen years later, Gauss stumbled upon Lobachevsky's article, and Wolfgang wrote him about his son's work, but Gauss wasn't about to publicize either of them and risk putting himself at the center of controversy. He wrote Bolyai a nice congratulatory letter (mentioning that he himself had already discovered similar results), and graciously proposed Lobachevsky as a corresponding member of the Royal Society of Sciences in Göttingen (he was immediately elected in 1842).

János Bolyai never published another work of mathematics. Lobachevsky became a successful administrator and eventually president of the University of Kazan. Bolyai and Lobachevsky might have both faded into the distant unknown, were it not for their contact with Gauss. Ironically, it was Gauss's death that finally led to the non-Euclidean revolution.

Gauss was a meticulous chronicler of things around him. He took pleasure in the collection of certain bizarre data, such as the length of life of deceased friends, in days, or the number of steps from the observatory at which he worked to various places he liked to visit. He also chronicled his work. After his death, scholars pored over his notes and correspondence. There, they discovered his research on non-Euclidean space, as well as the work of Bolyai and Lobachevsky. In 1867, both the Bolyais' and Lobachevsky's articles were included in the second edition of Richard Baltzer's influential *Elemente der Mathematik*. They soon became a standard reference among those working on the new geometries.

In 1868, the Italian mathematician Eugenio Beltrami laid to rest once and for all the issue of proving the parallel postulate: he proved that if Euclidean geometry forms a consistent mathematical structure, then so must the recently discovered non-Euclidean spaces. Is Euclidean geometry itself consistent? As we'll see, that has never been proved, or disproved.

17. Lost in Hyperbolic Space

HAT IS non-Euclidean space? The space that Gauss, Bolyai, and Lobachevsky discovered, hyperbolic space, is the space that results from replacing the parallel postulate by the assumption that for any line, there are not just one, but many parallel lines through any given external point. One consequence, as Gauss wrote Taurinus, is that the sum of the angles in a triangle is always less than 180 degrees, by a number Gauss called the *angular defect*. Another, the one stumbled upon by Wallis, is that similar triangles do not exist. The two are related, for the angular defect varies with the size of the triangle. Larger triangles have greater angular defects; smaller triangles are more nearly Euclidean. In hyperbolic space, the Euclidean form is approachable but not attainable, like the speed of light, or your ideal weight.

Though but a minor change in a simple axiom, the alteration of the parallel postulate produced a wave that propagated through the body of Euclidean theorems, changing each and every one that pertained to the shape of space. It was as if Gauss had removed the glass from Euclid's window and replaced it with a distorting lens.

Neither Gauss nor Lobachevsky nor Bolyai discovered any simple way of visualizing this new type of space. That was achieved by Eugenio Beltrami and, in simpler form, by Henri Poincaré, mathematician, physicist, philosopher, and first cousin of the future president of France, Raymond Poincaré. Then and now, Henri is the less famous Poincaré, but, like his cousin, Henri could turn a phrase. "Mathematicians are born, not made," Poincaré wrote. A cliché was born, and Henri's popular legacy secured. Less known outside academic circles is Henri's work in the 1880s, in which he defined a concrete model of hyperbolic space.

In creating his model, Poincaré replaced primitives like line and plane with concrete entities. He then interpreted the axioms of hyperbolic geometry in terms of them. It is acceptable to translate the undefined terms of space as curves or surfaces or even foods as long as the meanings they obtain from the postulates that apply to them are well defined and consistent. We could model the non-Euclidean plane as the surface of a zebra, call its hair follicles points and its stripes lines if we wanted, as long as it led to a consistent translation of the axioms. For instance, recall Euclid's first postulate as it applies to zebra space:

1. Given any two hair follicles, a segment of a stripe can be drawn with those follicles as its endpoints.

This postulate does not hold true in zebra space: a zebra's stripes have thickness and run only one way. Two hair follicles that are at equal positions along a stripe but are laterally displaced from one another are therefore not the endpoints of a segment of any stripe. There were no zebras in Poincaré's model. But it did resemble a crêpe.

Here is how Poincaré's universe works: the infinite plane is replaced by a finite disc, like a crêpe, but infinitely thin and with a perfectly circular boundary. The "points" are the things that have been considered points ever since Descartes: positions, like dots of fine white sugar. Poincaré's lines are like curved brown griddlemarks. In more technical terms, they are "any arcs of circles that intersect the boundary of the disc at right angles." In order to distinguish these lines from our intuitive picture of a line, we will call them "Poincaré-lines."

Having formed this physical picture, Poincaré had to give meaning to the geometric concepts that were to apply to it.

One crucial one was *congruence,* that pesky concept of equality of shape that Euclid prescribed you check by superimposing figures. As his fourth "common notion," Euclid wrote:

4. Things which coincide with one another are equal to one another.

As we have seen, the ability to move figures in space without distortion is guaranteed only if we assume the Euclidean form of the parallel postulate. So employing common notion 4 as a recipe for congruence is a no-no in non-Euclidean space. Poincaré's solution was to interpret congruence by defining a system of measurement for length and angle. Two figures would then be congruent if the lengths of their sides and the angles between them coincided. Seems obvious, right? But it isn't quite that simple.

Defining the measurement of angles was straightforward. Poincaré defined the angle between two Poincaré lines as the angle between their tangent lines at their point of intersection. To achieve a definition of length, or distance, Poincaré had to work harder. One might have expected some problems with this concept, since he was stuffing an infinite plane into a finite region. For instance, recall postulate 2:

2. Any line segment can be extended indefinitely in either direction.

Obviously, utilizing the usual definition of distance, this postulate would not hold true on the crêpe. But Poincaré redefined distance so that space is compressed as you approach the edge of his universe, effectively turning the finite area into an infinite one. It sounds easy, but Poincaré couldn't simply redefine distance at will—to be acceptable, his definition had to satisfy many demands. For instance, the distance between two different points must always be greater than zero. Also, the precise mathematical form Poincaré chose had to make the Poincaré-line that joins any two points the shortest path between them (called a *geodesic*), just as the usual line is the shortest path between points in Euclidean space.

If you examine all the fundamental geometric concepts needed to define hyperbolic space, you'll find that Poincaré's model leads to a consistent interpretation for each one. We can verify the others, but the most interesting one to look at is the famous parallel postulate. The hyperbolic version of the parallel postulate, given here for the Poincaré model in Playfair's form, reads:

> Given a Poincaré-line and a point not on that Poincaré-line, there are many other Poincaré-lines that pass through the external point and do not intersect the given Poincaré-line.

The figure on the following page illustrates how this is possible.

Poincaré's model of hyperbolic space is a laboratory that makes it easy to see some of the unusual theorems and properties mathematicians previously worked so hard to uncover. For example, suppose we attempt to draw a rectangle, which does not exist in a non-Euclidean space. First, draw a Poincaré-baseline. Then, on the same side of the baseline, draw two Poincaré-line segments that are perpendicular to it. Finally, connect the two segments with another segment that, like the baseline, is perpendicular to both of them. It's impossible. There are no rectangles in Poincaré's world.

What did Poincaré accomplish with all this? One might imagine a few bespectacled mathematicians at the University of Paris politely applauding smarty-pants Henri after a seminar on his model. Perhaps inviting him for an after-lecture ab-

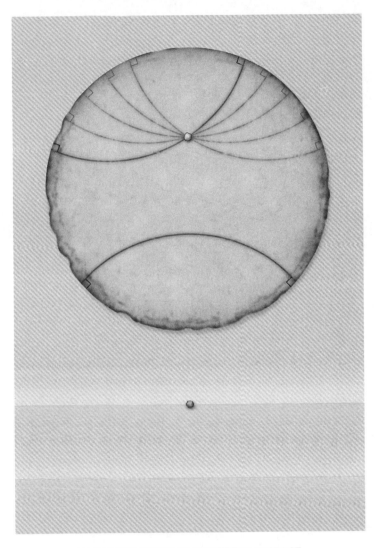

PARALLEL LINES IN HYPERBOLIC SPACE
AND EUCLIDEAN SPACE

sinthe or maybe a crêpe on which they could draw rectangles in the jam. But why would anyone, over a century later, be writing a book on this stuff, or why would you, an intelligent and busy reader with many other things to do, be reading it?

Here is the punch line: Poincaré's model isn't just a model of hyperbolic space, it *is* hyperbolic space (in two dimensions). In the language of mathematics, this means mathematicians have proven that all possible mathematical descriptions of the hyperbolic plane are isomorphic—the mathematicians' way of saying that they are the same. If our space is hyperbolic, it will behave exactly like Poincaré's model (except in three dimensions). To paraphrase the Disney song, it's a small crêpe after all.

● ■ ▲

A couple decades after the discovery of hyperbolic space, another type of non-Euclidean space was discovered: elliptic space. Elliptic space is the space you get if you assume the other violation of the parallel postulate: that no parallel lines exist (i.e., all lines in the plane must intersect). In two dimensions, this type of space was known and studied in a different context by the Greeks, and even by Gauss, yet they didn't grasp its significance as an example of an elliptic space. And for good reason: it had been proved that in Euclid's system, even if one allowed for alternate forms of the parallel postulate, elliptic spaces cannot exist. In the end, though, it was not elliptic spaces that proved problematic, it was Euclid's axiomatic structure itself.

18. Some Insects Called the Human Race

N THE DECADE starting with 1816, Gauss spent a lot of time away from home directing a massive effort to survey areas of Germany, an endeavor we today would call a geodetic survey. The point of the survey was to measure the distance between cities and other landmarks, and to put this data together into a map. The exercise is not as easy as it may seem, for a couple reasons.

The first difficulty Gauss had to overcome was that the surveying instruments had limited range. Because of this, straight lines had to be constructed from shorter segments, each of which had a certain degree of random error of measurement. The errors added up quickly. Gauss did not respond to this difficulty in the manner of your normal researcher, say, the author of this book. That would involve, first, a lot of pulling at his hair and occasional snapping at his children; second, achievement of some tiny, incremental progress; and third, publication of the result, phrasing it in a manner to make it seem as important as possible. Instead, Gauss invented the central concept of the modern field of probability and statistics—the theorem that random errors will be distributed in a bell-shaped curve around a mean.

Having put the error problem behind him, Gauss faced the challenge of patching together a two-dimensional map from three-dimensional data affected by variations in elevation as well as the curvature of the earth. The difficulty arises from the fact that the surface of a globe just does not have the same geometry as the Euclidean plane. It is the mathematician's version of the quandary faced by any parent who has ever tried to wrap a round ball with flat gift wrap. If, as the parent, you overcame the difficulty by cutting the paper into tiny squares and patching them together onto the ball, then you solved the problem the same way Gauss did, minus the technical details. Those details, Gauss published in a paper in 1827. Today, a whole field of mathematics has grown up around it, a field called differential geometry.

Differential geometry is the theory of curved surfaces in which a surface is described by the coordinate method invented by Descartes, and then analyzed employing differential calculus. It sounds like a narrow theory, applicable perhaps to coffee mugs, airplane wings, or your nose, but not to the structure of our universe. Gauss had other ideas. In his paper, he made two key realizations. First, he asserted that a surface could be considered a space in itself. We could, for instance, think of the surface of the earth as a space, which, in our daily lives, is certainly the role it played before the advent of air travel. It's probably not the type of thing Keats had in mind when he wrote of "the universe in a grain of sand," but the poetry fits the mathematics.

The other breakthrough idea Gauss established was that the curvature of a given space could be studied solely on the surface itself, without reference to a larger space that may— or may not—contain it. More technically, the geometry of a curved surface can be studied *without reference to a higher-dimensional Euclidean space.* That a space could "curve" yet not be curving into anything was a concept that would later prove necessary in Einstein's general theory of relativity. After all, since we cannot step out of our universe to gaze down at our limited three-dimensional realm, only this sort of theorem can give us hope of determining the curvature of our own space.

To understand how we can detect curvature without employing an imbedding space, imagine Alexei and Nicolai now as two-dimensional beings in a civilization strictly confined to the space that is the earth's surface. How are their experiences different from ours, apart from no airplane trips, no climbs of Mount Everest, and the fact that their record for the Olympic high jump would be zero?

Take the high jump record. It is not just that Alexei couldn't get off the ground. To him, the *concept* of getting off the ground would not exist. This is no reason for we 3-D-ites to feel superior. At this moment, at a dinner party of four-dimensional beings, a few amused souls may very well be sipping Margaritas as they gaze "down" at us and muse at our own limitations. Like a race of crawling insects, of jumping "up" in their 4-D space we sorry creatures possess no concept.

To say that Alexei and Nicolai couldn't climb Mount Everest also requires clarification. They certainly could reach the top—after all, it is a part of the earth's surface. But they would not have a concept of change in elevation. As Alexei left the base of the mountain and walked toward its peak, what we know as gravity would appear as a mysterious force pushing him back toward the base—as if the mountain peak possessed some strange repulsive quality.

Along with the mysterious force would come a distortion of the geometry of space. Any triangle, for instance, that contained the mountain in its interior, would enclose a mysteriously large area. We can understand this because the surface of the mountain is greater than the area of its footprint, but to Alexei and Nicolai it would represent a warping of space.

Alexei and Nicolai could imagine no sticks protruding from the sand and observe no suns casting shadows from out in space. A boat disappearing on the horizon would be flat, not differentiated into hull and masts. All the hints the ancients drew upon to discern the roundness of our planet would vanish, and all they would know would be the distances and relationships between points in their space. With-

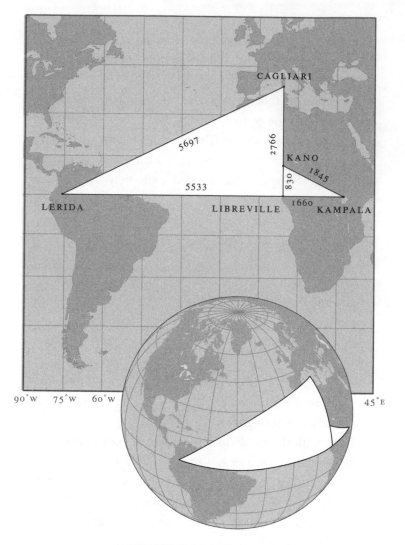

TRIANGLES ON THE GLOBE

out hints from the third dimension, Euclid himself might have concluded that space was non-Euclidean.

Imagine in this world an ancient scholar named Noneuclid. Sitting in her office at the academy she has drawn the same conclusions as our man Euclid. But before she publishes her

Elements, she wants to check that her theories apply beyond her walls, to the large-scale geometry of space. Her graduate student Alexei brings her a map from the library. It is the map shown.

The map shows Libreville, Gabon, located at 00° latitude, 09° East longitude at the vertex of a right triangle. Moving north 12 degrees, which puts you roughly at Kano, Nigeria, and east 24 degrees, which lands you at Kampala, Uganda, you trace the base legs of the triangle. One of the basic theorems of Euclidean geometry is the Pythagorean theorem. Noneuclid asks Alexei to do the math to check it. Alexei reports back:

Sum of squares of bases:	3,444,500
Square of hypotenuse:	3,404,025

Noneuclid, on seeing this data, growls at Alexei's sloppiness. Redoing the calculation, though, Noneuclid finds that Alexei was right. Noneuclid now adopts the next line of defense of the theorist—she ascribes the discrepancy to experimental error. She sends her other student, Nicolai, back to the library for more data. Nicolai returns with an even larger triangle formed by Libreville; Cagliari, Italy, at 39° North; and Lerida, Colombia, at 71° West. This triangle is also shown on the map. Nicolai finds:

Sum of squares of bases:	38,264,845
Square of hypotenuse:	32,455,809

Noneuclid is not pleased. This discrepancy looks even worse. How could her colleague Nonpythagoras have been so wrong? How could Noneuclid have measured dozens of triangles and never noticed a problem? Those, chimes in Alexei, were tiny triangles; these are huge. Nicolai notices that the discrepancy is larger for larger triangles. He hypothesizes that all previous triangles studied, measured in their tiny laboratory, or around town, were so small that the deviation went unnoticed.

Noneuclid decides to use some grant money to send Alexei and Nicolai on an expedition to New York. Starting there, 40° 45" N latitude, 74°00" W longitude, she has instructed Alexei to walk ten minutes (of longitude) due west, which brings him to, roughly, downtown Newark. Nicolai walks ten minutes (of latitude) due north, which brings him to New Milford, New Jersey. To a good accuracy, the three points form a right triangle, with these legs: New York to Newark, 8.73 miles; New York to New Milford, 11.53 miles; and New Milford to Newark, 14.46 miles.

Noneuclid checks the Pythagorean theorem:

Sum of squares of bases:	209
Hypotenuse (Newark to New Milford):	209

For small enough triangles, it works. As the beginnings of non-Euclidean geometry percolate in her head, Noneuclid sends her students off on one last expedition.

This time, Alexei and Nicolai are to sail between New York and Madrid, which at 40°N, 04°W, is almost due west of New York. They are not to sail once, but many times between the two cities, with slightly different routes each time, each time measuring the precise length of their path. Their search, like Columbus's, is for a shorter route between lands. In Alexei and Nicolai's case, it is for the geodesic, or shortest route. It is several years' work, but the resulting publication would have made a big splash.

Is the shortest route, from New York to Madrid, to sail "straight" east along their common line of latitude? No. Instead, it is to sail along the strange curved line as shown on the map, to first head northwest, then gradually turn your

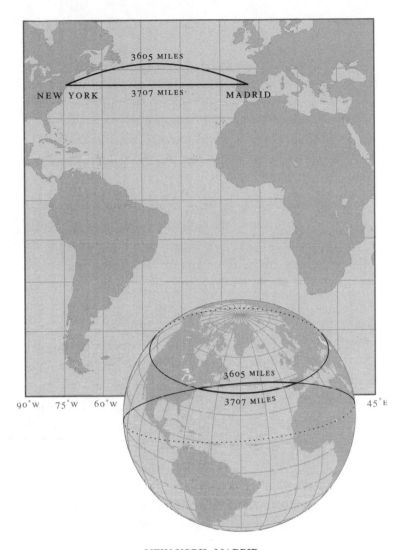

NEW YORK—MADRID

heading more southerly until you are moving southwest. It is the same route a bowling ball would follow if you could roll it unimpeded, or that certain genius birds, such as American Golden Plovers and Bristle-thighed Curlews, follow in their migration. It is also the route 2-D Egyptian rope stretchers marked off as they pulled their ropes taut from point to point.

This is easy to understand if you visualize the earth as viewed from space. Heading straight east doesn't work because as you travel along the globe, the directions called "north" and "east" are not fixed directions. As you move from New York toward Madrid, the direction called east rotates in three-dimensional space, as does the one called north. The shortest route between New York and Madrid, or between any two points on the globe, is along a curve called a great circle (a circle on the globe whose center coincides with the center of the earth; they are the largest circles you can draw on the earth's surface, hence their name). The great circles are the analogues of the Poincaré-lines in Poincaré's universe, the curves you would naturally call lines, and that play the role of lines in Euclid's axioms. Lines of longitude are great circles. So is the equator, but it is the only curve of constant latitude that is (all the other circles of constant latitude have their centers further up or down the earth's axis).

The view from outer space is not the view of the indigenous people such as Noneuclid. To her, there is no "center of the earth," and, as Gauss showed was possible, there is no "outer space." Inspired by Alexei and Nicolai's measurements, she would conclude that the space in which she lives is a non-Euclidean space; not hyperbolic space, but the space appropriate to the surface of a globe: elliptic space.

In Noneuclid's space, all lines, the great circles, intersect. Angle sums in triangles are all *greater* than 180° (in hyperbolic space they are *less*). A triangle formed by the equator and two lines of longitude connecting the equator to the North Pole, for example, would have an angle sum of 270°. As in hyperbolic space, this space would look more Euclidean at smaller distances, which is why it took so long to notice the deviation. For instance, the number of degrees of angle sum in excess of 180° shrinks as a triangle gets smaller.

The geometry of elliptic spaces—called spherical geometry—was well known even in antiquity. The great circles were known to be the geodesics. Geometric formulae relating the parts of spherical triangles had been discovered and applied to mapmaking. But elliptic spaces did not fit into Euclid's paradigm, and the discovery that the globe is an elliptic space was left for one of Gauss's students, Georg Friedrich Bernhard Riemann. It was made in Gauss's declining years, but this discovery more than any other eventually fueled the curved space revolution.

19. A Tale of Two Aliens

EORG RIEMANN was born in 1826 in the small village of Breselenz near Gauss's birthplace. He had five siblings. Most of them, as he was destined to do, died young. His mother died before he was grown. He was schooled at home by his father, a Lutheran pastor, until he was ten. His favorite subject was history, particularly the Polish national movement. If earnest little Georg doesn't sound like the life of the party, he wasn't. In fact, he was pathologically shy and modest. And brilliant. On the evidence of Gauss and Riemann, a conspiracy theorist might conclude that around the turn of the nineteenth century, a superior alien race formed a colony near Hannover, Germany, and deposited at least two genius infants with poor families in the area. Though there are no stories of Riemann as a toddler genius like Gauss, Riemann too seemed a bit too smart to be one of us.

When Riemann was nineteen, the director of his Gymnasium, a man named Schmalfuss, gave him a little something to look over: Adrien-Marie Legendre's book, *Théorie des nombres,* or *Theory of Numbers*. It was the mathematics equivalent of lending young Riemann the barbell for the world record in the standing chest press. This barbell weighed in at 859 pages—large, dense pages, crammed with abstract theory. It was hernia material, something only a champion could master, and only with a lot of sweat and grunting. To Riemann, the book was a lightweight, a page-turner apparently requiring no strenuous concentration. He returned it in six days with a comment along the lines of "It was a good read." Several months later he was examined on its contents and achieved a perfect score. Later still he would make his own fundamental contributions to number theory.

In 1846, Riemann, still nineteen, matriculated at the University of Göttingen, where Gauss was a professor. Riemann began as a student of theology, perhaps so he could pray for the downtrodden Poles. He soon switched to what had become his first love, mathematics. After a brief period in Berlin, Riemann returned to Göttingen in 1849 to complete his Ph.D. His dissertation, submitted in 1851, was reviewed, among others, by Gauss, who was by then a legend, and legendarily tough on students.

Gauss's reaction to Riemann's work followed the familiar pattern Gauss took on the rare occasions he was impressed with a piece of mathematics. Gauss wrote that Riemann had demonstrated "a creative, active, truly mathematical mind, and . . . a gloriously fertile imagination," and also remarked that he, Gauss, had earlier done similar work, but not published it. (Posthumous examination of Gauss's papers later showed that all his claims were true.) Riemann was delighted. By 1853, Riemann was twenty-seven and on the last stretch of the long road to a lectureship at Göttingen. In Germany at that time, such an academic position did not pay the modest salary it does today. It did not pay *any* salary. To many of us, that would be a bit of a drawback. To Riemann, it was a coveted position, a stepping stone to a professorship. And stu dents gave tips.

Riemann's final hurdle was to give a trial lecture. He submitted three topics from which the faculty could choose. It was customary for them to choose the applicant's first topic. Just in case, Riemann was well prepared for either his first or his second. Gauss, always a barrel of laughs, chose Riemann's third.

For his third choice, Riemann had selected a topic in which

he must have had some interest, but of which he had little knowledge. Most academics interviewing for a job, if their research was on the politics of Luxembourg, would not propose to give their audition colloquium on the reptiles of Sri Lanka, even if it was the third down on their list. When Gauss, by then seriously ill and told by his doctor that death was imminent, chose Riemann's third topic, Riemann must have asked himself, "What was I thinking?" The topic that he had listed was *Über die Hypothesen welche der Geometrie zu grunde liegen* ("On the Hypotheses Which Lie at the Foundations of Geometry"). He had chosen a topic that he knew had been close to Gauss's heart for virtually his entire life.

Riemann's next step was understandable—he spent several weeks having some kind of breakdown, staring at the walls, paralyzed by the pressure. Finally, when spring came, he pulled himself together and in seven weeks hammered out a lecture. He presented it on June 10, 1854. It proved to be one of the rare occasions in history that the exact date and details of a job interview have been preserved for posterity.

Riemann phrased his lecture in the context of differential geometry, focusing on the properties of infinitesimally small regions of a surface rather than its gross geometric characteristics. In fact, Riemann never mentioned non-Euclidean geometry by name. But the implications of his work were clear: Riemann explained how the sphere could be interpreted as a two-dimensional elliptic space.

Like Poincaré, Riemann gave his own interpretations of the terms *point, line,* and *plane.* As the plane, he chose the surface of the sphere. His points, like Poincaré's, were positions, in the manner of Descartes considered to be pairs of numbers, or coordinates (essentially the point's latitude and longitude). Riemann's lines were the great circles, the geodesics on the sphere.

As in Poincaré's model, it must be confirmed that Riemann's admits consistent interpretations of the postulates. Now might be a good time to recall that it had been proved that elliptic space could not exist. Sure enough, Riemann's model did turn out to have a few little problems. It was one thing to create a space based on a new version of the parallel postulate; Riemann's space was inconsistent with the existing versions of other postulates as well. For instance, consider postulate 2. Euclid wrote:

2. Any line segment can be extended indefinitely in either direction.

Is this satisfied by segments of great circles on the sphere? Before Riemann, postulate 2 was interpreted as meaning that there must exist line segments of arbitrarily large length. But there is a limit to the length of a great circle, its circumference, 2π times the radius of the sphere.

Even in mathematics, it sometimes pays to break the law. Here Riemann was Rosa Parks refusing to sit at the back of the bus, questioning, if not the unjust, the unjustified. He asserted that the second postulate was needed, not to make line segments arbitrarily long, but only to guarantee that lines had no bounds, which *is* true of the great circles. In mathematics the Supreme Court is the community of mathematicians, and they scratched their heads at this. What are the implications of young Riemann's new interpretation of the law? Is it consistent with other laws? Can it be made consistent?

Actually, the contradictions didn't stop with postulate 2. Riemann's concept of line led to other problems for which Riemann offered no explanation. For instance, great circles violate the assumption that two lines can intersect at only one point. Like lines of longitude that intersect at both the North and South poles, all great circles intersect at two points, on opposite sides of the sphere.

The concept of inbetweenness also became hard to interpret. Euclid based the concept of inbetweenness on postulate 1:

1. Given any two points, a line segment can be drawn with those points as its endpoints.

To produce a point between two given points, Euclid would draw the segment connecting the two. Any point (other than the endpoints) on this segment is considered "between" the other two. The problem in Riemann's model is that there are always two ways to connect a pair of points along a circle. Is Indonesia between equatorial Africa and equatorial South America? To decide, you trace a line along the equator connecting the two continents and check whether it passes through Indonesia. But in Riemann's model, you can get from South America to Africa by traveling either east or west. One path passes through Indonesia, the other does not.

Due to this ambiguity, on the globe all Euclid's proofs that involve constructions connecting points with line segments become ill-defined. This leads to some bizarre consequences. For instance, imagine Riemann's spherical universe with a radius of 40 miles, instead of the earth's 4,000. On a clear day, you might look ahead and see your behind. Is your backside behind you or in front of you? Or take the hula hoop. Its radius is about 1 meter. Wiggling the hoop about your waist, you ask, are you inside it? It sure seems so. Now imagine expanding the hoop. Blow it up to the size of a race track, a mile wide. Large for a hula hoop, but still small compared to the planet's 40-mile radius. Standing in the middle, you still feel safe declaring you are inside the hoop. Now stretch the hoop to a radius of 40 miles. It exactly encircles the planet, like an equator, and suddenly it seems arbitrary whether you consider yourself inside the hoop or on the outside. Expand its radius further, that is, push its circumference further from you, and the hoop actually *shrinks*. Eventually, it looks just as it did when you started—a meter in radius, but now centered at a point across the world from you. You seem to be on the outside. How can you pass from inside to outside by merely expanding the hoop? With the demise of inbetween, behind and in front, inside and outside, are no longer simple concepts. These are the contradictions of naive elliptic space.

Eliminating these quandaries involves the careful redefinition of many concepts. As usual, Gauss had foreseen this. He had written to Wolfgang Bolyai in 1832, "In a complete development such words as 'between' must be founded on clear concepts, which can be done, but which I have not found anywhere." Nor did he get them from Riemann. But, concentrating mainly on small regions of the surface, global contradictions such as we have described seemed to neither deter nor interest Riemann. And, despite these open issues, Riemann's lecture is considered one of the great masterpieces of mathematics. Still, with all these loose ends, it did not immediately illuminate like a photon torpedo the universe of mathematicians. Gauss died soon after Riemann's lecture; Riemann continued to focus on questions of local structure rather than the large-scale geometry of space, and his work had no great impact in his lifetime.

In 1857, at age thirty-one, Riemann finally obtained an assistant professorship, with a meager salary equivalent to about $300 a year. On this, Riemann supported himself and his three surviving siblings, all sisters, though the youngest, Marie, died soon after. In 1859, Gauss's successor Dirichlet died and Riemann was promoted to Gauss's spot on the faculty. Three years later, at age thirty-six, he married. The next year he had a baby girl. With a decent salary and the beginnings of a family, things were looking up for Riemann. But it wasn't to be. He contracted pleurisy, which turned into tuberculosis, and felled him like his siblings at an early age—thirty-nine.

Riemann's work on differential geometry became the cornerstone of Einstein's general theory of relativity. Had Riemann not been so imprudent as to include geometry on his topic list, or Gauss not so bold as to choose it, the mathematical apparatus Einstein needed for his revolution in physics would not have existed. But before that upheaval would begin, Riemann's work on elliptic spaces had an equally profound impact on the world of mathematics. The need to alter postulates other than the parallel postulate was like the fraying of the strands in a rope. Soon, the rope snapped. Only then did mathematicians realize that hanging from it had been not only geometry, but all of mathematics.

20. After 2,000 Years, A Face-lift

IEMANN'S 1854 lecture was not published until 1868, two years after his death, and one year after Baltzer's book popularized Bolyai's and Lobachevsky's work. Gradually, the implications of Riemann's work demonstrated that Euclid had made several types of errors: he had made many unstated assumptions; he had made other assumptions that were not formulated properly; and he had attempted to define more than was possible.

Today, we see many faults in Euclid's reasoning. One easy criticism of Euclid is his artificial separation of postulates from "common notions." The deeper point here is that today we seek to axiomatize all our assumptions, and accept nothing as truth merely on the basis of "reality" or "common sense." This is a fairly modern attitude, a victory of Gauss over Kant, and it is hard to criticize Euclid for not making such a leap.

Another structural problem in Euclid's system was not recognizing the need for undefined terms. Consider the dictionary definition of space as "unlimited room or place extending in all directions." Is this a meaningful definition, or have we just substituted the vague term *place* for our object word *space?* If we don't feel we understand the term *place* precisely enough, we can, of course, look it up, as well. The dictionary states that "place" is "the part of space occupied by a certain object." The two words *place* and *space* are often defined in terms of each other.

It may take a bit longer, but since every word in a dictionary is defined in terms of another word, this has to happen eventually with any definition. The only way to avoid circular reasoning in a finite language would be to include some undefined terms in the dictionary. Today we realize that mathematical systems, too, must include undefined terms, and seek to include the minimum number necessary for the system to make sense.

Undefined terms must be handled with care, for we can easily be led astray if we read meaning into a term without first proving it, even if the meaning seems obvious from our physical picture. Thābit made this error when he used the intuitively "obvious" property that a curve everywhere equidistant to a line is also a line. As we saw, there is nothing in Euclid's system, outside the parallel postulate itself, that guarantees this. When we employ undefined terms, we must ignore all connotations the choice of words seems to imply. To paraphrase the great Göttingen mathematician David Hilbert, "One must be able to say at all times—instead of points, lines, and circles—men, women, and beer mugs."

An undefined term does not long remain devoid of meaning: it receives definition from the postulates and theorems that apply to it. For instance, suppose, as Hilbert mused, we changed the names of the undefined terms *point, line,* and *circle* to *man, woman,* and *beer mug.* Then, mathematically, these terms would obtain meaning from statements such as these, Euclid's first three postulates:

1. Given any two men, a woman can be drawn with those men as its endpoints.

2. Any woman can be extended indefinitely in either direction.

3. Given any man, a beer mug with any radius can be drawn with that man at its center.

Euclid made other errors, of pure logic, leading him to prove some theorems employing steps that are unjustified.

For example, in his very first proposition he claims to show that an equilateral triangle can be constructed on any given line segment. In his proof he forms two circles, one centered at each end of the line segment, and each with a radius equal to the length of the segment. Then he makes use of the point where the two circles intersect. Though drawing the circles will clearly show this intersection, he includes nothing in his formal argument to guarantee the existence of this point. In fact, his system lacks a postulate that would ensure continuity of lines or circles, that is, ensure that there are no gaps in them. He also failed to recognize other assumptions he frequently used in proofs, such as the assumption that points and lines exist, that not all points are collinear, and that every line has at least two points on it.

In another proof, he tacitly assumed that if three points lie on the same line, we can identify one of the points as being between the other two. Nothing in his postulates or definitions allows you to prove this. In reality, this assumption is really a kind of straightness requirement: it disallows lines that curve because such lines could form a closed loop like a circle, and then you could not identify any of the points as the one in between the other two.

Some of the objections to Euclid's proofs might seem like nitpicking, but innocent, obvious assumptions of no apparent consequence can sometimes be equivalent to major theoretical statements. For instance, assuming the existence of just one triangle whose angles sum to 180 degrees allows one to prove that all triangles have angle sums of 180 degrees, and also allows us to prove the parallel postulate.

In 1871, the Prussian mathematician Felix Klein showed how to rectify the apparent contradictions in Riemann's sphere model of elliptic space, improving on Euclid in the process. Mathematicians like Beltrami and Poincaré soon suggested their new models and new approaches to geometry.

In 1894, Italian logician Giuseppe Peano proposed a new set of axioms to define Euclidean geometry. In 1899, Hilbert, who did not know of Peano's work, gave his first version of the formulation of geometry that is most accepted today.

Hilbert was utterly dedicated to clarifying the foundations of geometry (and later also helped develop Einstein's general theory of relativity). He revised his formulation many times before his death in 1943. The first step in his method was to turn Euclid's unstated assumptions into explicit statements. In Hilbert's system, at least in the seventh edition of his work, published in 1930, he included eight undefined terms and increased the number of axioms from Euclid's ten (including common notions) to twenty. Hilbert's axioms were divided into four groups. They include assumptions unrecognized by Euclid such as these we have already considered:

Axiom I-3: There are at least two points on each line. There exist at least three points in space which are not all on the same line.

Axiom II-3: Given any three points on a line, only one of them can lie between the other two.

Hilbert and others showed that all properties of Euclidean space follow from his axioms.

● ■ ▲

The curved space revolution had a profound influence on all areas of mathematics. From around Euclid's time until the time Gauss and Riemann's work was posthumously discovered, mathematics was largely pragmatic. Euclid's structure was interpreted as describing physical space. Mathematics was, in a sense, a form of physics. Questions of the consistency of mathematical theories seemed moot—the proof was in the physical world. But by 1900, mathematicians had the view that axioms were arbitrary statements, merely the basis of a system whose consequences would be investigated in a kind of mental game. Suddenly, mathematical spaces were considered abstract logical structures. The nature of physical space became a separate issue, a question of physics, not mathematics.

For mathematicians, a new type of question now arose: the question of proving the logical consistency of their structures. The idea of proof, which took a back seat during recent centuries of advances in calculational technique, again became dominant. Is Euclid's geometry self-consistent? The most straightforward way to prove the consistency of a logical system is to prove all possible theorems and show that none contradict each other. Since there are an infinite number of possible theorems, this is an intelligent approach only for those of us who plan to live forever. Hilbert tried another tactic. Like Descartes and Riemann, Hilbert identified points in space with numbers. In the case of two-dimensional space, for instance, each point corresponds to a pair of real numbers. By turning points into numbers, Hilbert was able to translate all fundamental geometric concepts and axioms into arithmetic ones. Thus, the proof of any geometric theorem translates to an arithmetic or algebraic manipulation of coordinates. And since any geometric proof follows logically from the axioms, the arithmetic interpretation must also follow logically from the axioms in arithmetic form. If any contradiction were to arise from the geometry, it would translate into a contradiction in arithmetic: if arithmetic is consistent, then so is Hilbert's formulation of Euclidean geometry (this was also eventually done for the non-Euclidean geometries). Plain as day? The bottom line is, though Hilbert did not show the *absolute* consistency of geometry, he did show what is called its *relative* consistency.

Due to the infinity of possible theorems, the absolute consistency of geometry, arithmetic, and for that matter all of mathematics, is a harder question. To get at it, mathematicians invented an abstract theory of objects that deals with them only on the most general level, independent of the specific nuances, and nuisances, related to what they really are. This theory, now taught in some form in most grade schools, is called set theory.

Yet even simple set theory faces perplexing paradoxes, such as this famous one published in the obscure journal *Abhandlung der Friesschen Schule* by Kurt Grelling and Leonard Nelson in 1908. Grelling and Nelson consider sets of words. First, the set of all adjectives that describe the words themselves. For example, the word "twelve-letter" is, yes, a twelve-letter word, and the adjective "polysyllabic" is polysyllabic. Standing opposed to this set is the set of all adjectives that do *not* describe themselves. For some reason, adjectives like "well written," "fascinating," and "recommendable-to-a-friend" come to mind (if there is one sentence from this book that should be committed to memory, it is this one). This latter set of words is usually called *heterological,* perhaps because heterological is polysyllabic.

So far, so good. But here's the catch: is heterological a heterological word? If it is, then it describes itself, so it is not. If it is not, then it does not describe itself, so it is. Mathematicians call this a paradox; to non-mathematicians, it is just the familiar lose-lose situation (a term that originated with mathematicians, bless them).

● ■ ▲

In 1903, in an effort to clean up the field, Bertrand Russell, soon to be Lord Russell, suggested in a modest book entitled *Principles of Mathematics* that all mathematics should be derivable from logic. He attempted to accomplish the derivation, or at least to show how to do it, with his colleague at Oxford Alfred North Whitehead in a three-volume magnum opus published between 1910 and 1913. Presumably because it was more serious than the 1903 version, it received a Latin title, *Principia Mathematica.* In *Principia,* Russell and Whitehead claimed to have reduced all of mathematics to a unified system of basic axioms from which all theorems of mathematics could be proved, just as Euclid had attempted to do for geometry. In their system, even entities as fundamental as numbers were considered empirical constructs that had to be justified by a deeper, more fundamental axiomatic structure.

Hilbert was skeptical. He challanged mathematicians to prove rigorously that the program of Russell and Whitehead succeeded. This question was settled for good in 1931 by the shocking theorem of Kurt Gödel: he proved that in a system of sufficient complexity, such as the theory of numbers, there must exist a statement that cannot be proved either true or false. A corollary of Gödel's theorem is that there must exist a true statement that cannot be proved. This destroys the claims of Russell and Whitehead—not only did they not show how all mathematical theorems can be derived from logic, it is actually impossible to do so!

Mathematicians continue to work on the foundations of their field, but none of the developments since Gödel have changed the picture very much. There is still no universally accepted approach to what Euclid had begun—the axiomatization of mathematics.

Meanwhile, the power of mathematics as more than a mental game is nowhere more evident than in the application by Einstein of the newly discovered types of mathematical space to the description of the space in which we live. Though thoroughly remodeled, geometry continued to be the window to understanding our universe.

IV

THE STORY

OF EINSTEIN

What makes space curve? Space is given a new dimension as space-time explodes on to the twentieth century and makes a Patent Office clerk the hero of the century.

21. Revolution at the Speed of Light

AUSS and Riemann showed that space could be curved, and gave the mathematics needed to describe it. The next question is, what kind of space do we live in? And, probing deeper: what determines the shape of space?

The answer, given so elegantly and precisely by Einstein in 1915, was actually first proposed in 1854, in broad strokes, by Riemann himself:

> The question of the validity of geometry . . . is related to the question of the internal basis of metric [distance] relationships of the space . . . we must seek the ground of its metric relations outside it, in the binding forces which act on it. . . .

What makes things far apart or close together? Riemann was too far ahead of his time to be able to develop a concrete theory based upon his insight, too far ahead even for his words to be appreciated. Sixteen years later, though, one mathematician did take notice.

On February 21, 1870, William Kingdon Clifford presented a paper to the Cambridge Philosophical Society entitled "On the Space Theory of Matter." Clifford was twenty-five that year, the same age as Einstein when he published his first articles on special relativity. In his paper, Clifford boldly proclaimed,

> I hold in fact: (1) That small portions of space are of a nature analogous to little hills on a surface which is on the average flat. (2) That the property of being curved or distorted

153

is continually passed on from one portion of space to another after the manner of a wave. (3) That this variation of the curvature of space is really what happens in that phenomenon which we call the motion of matter. . . .

Clifford's conclusions went far beyond Riemann's in their specificity. Which would hardly be notable except for one thing: he got it right. The reaction of a physicist reading this today has got to be, "How did he know?" Einstein came to similar conclusions only after years of careful reasoning. Clifford didn't even have a theory. However, Clifford managed to intuit such detailed conclusions, he, Riemann, and Einstein were all guided by the same simple mathematical idea: if objects in free motion move in the straight lines characteristic of Euclidean space, then might not other kinds of motion be accounted for by the curvature of non-Euclidean space? And in the end, it was precisely Einstein's careful reasoning, based on *physics,* not mathematics, that enabled him to develop the theory that Clifford could not.

Clifford worked feverishly on his theory, usually all through the night, for the day was too burdened with teaching and administrative duties at University College London. But without the deep understanding of physics that led Einstein to the intermediate step of special relativity, and the proper role of time, Clifford had little chance of developing his ideas into a workable theory. The mathematics had preceded the physics—a difficult situation, reminiscent, as we'll see, of the state of string theory today. Clifford got nowhere. He died in 1876, some say of exhaustion, at age thirty-three.

One problem Clifford had was that he found himself leading a parade of one. In the world of physics, the sky was sunny and bright, and few saw reason to spend their time attacking laws in which they saw no sign of corruption. For over 200 years it had seemed that every event in the universe was explicable by Newtonian mechanics, the theory based on the ideas of Isaac Newton. In Newton's view, space is "absolute," a fixed, God-given framework upon which to lay the coordinates of Descartes. The path of an object is a line or other curve described by a set of numbers, the coordinates that label the points the path covers in space. The role of time is "to paramatrize" the path, mathematician's lingo for "to tell you where you are along it." For instance, if Alexei is walking up Fifth Avenue at a steady speed of one block per minute, starting at 42nd Street, then his position is simply Fifth Avenue and (42 plus the number of minutes)nd Street. By specifying the number of minutes he has walked, you are determining where along the path he is.

With this understanding of time and space, Newton's laws predict how and why an object like Alexei moves—they give his position as a function of the parameter called time. (This of course assumes he is an inanimate object, which is only true some of the time; picture him with Discman earphones on.) According to Newton, Alexei will continue in uniform motion—in a straight line *and* at a constant speed—unless acted upon by an external force, such as the attraction of a video game arcade around the corner. Or, given such an attraction, Newton's laws predict how Alexei's path will differ from uniform motion. They will tell you, quantitatively, exactly how he will move, given his personal inertia and the strength and direction of the force. According to these equations, a body's acceleration (which is change in speed *or* direction) is proportional to the force applied to it and inversely proportional to its mass. But the description of the motion of a body reacting to a force is only half the picture, known as the "kinetics." To form a complete theory, we also need to know the "dynamics," that is, how to determine the strength and direction of the force, given the source (the arcade), the target (Alexei), and their separation. Newton gave such a force equation for only one type of force, the gravitational force.

Putting the two sets of equations together, the force equations (dynamics) and the motion equations (kinetics), one could (in principle) solve for an object's path as a function of time. One could predict, say, Alexei's orbit around the arcade, or (sadly) the path of a ballistic missile flying between two continents. Newton had fulfilled the ambition, which had begun with Pythagoras, to create a system of mathematics that permits the description of motion. And, by explaining how the same law governs motion on earth and in space, Newton did something else that was equally important: he united two old and separate disciplines—physics, which had been thought of as primarily concerned with everyday human experience, and astronomy, which had been concerned with the motion of heavenly bodies.

● ■ ▲

If Newton's view of space and time is true, then it is easy to see two things that cannot be. First, there can be no limit to the speed at which one thing can approach another. To see this, imagine that there is such a limiting speed; call it c. Next, imagine that an object is approaching you at that speed. Now (for the sake of science) spit at the object. If this drama occurs in a tangible medium called absolute space, it is easy to see that the object is now approaching your saliva faster than it is approaching you. The speed limit law is violated. Second, the speed of light cannot be constant. More precisely, light must approach different observers at different speeds. If you race toward light, it will approach you faster than if you run from it.

If an objective structure for space exists, these two truths are self-evident. Yet these two "truths" are false. This is the basis of special relativity, the ingredient missing from earlier speculations on the physics of curved space. It is a fact that was "observed" long before it was "appreciated."

22. Relativity's Other Albert

FEW YEARS after young Riemann took such a keen interest in Polish history, a young couple from the ethnic Polish province of Posnan, then under Prussian rule, had a baby boy they named Albert. One imagines that the heroic struggle of Polish nationalism was more attractive as reading material than real life experience. And if the Polish were heroes, they were also anti-heroes, exhibiting a virulent anti-Semitism that later made Poland Hitler's country of choice when locating his gas chambers. For whatever reason, around 1855, the year Gauss died, Albert's family, the Michelsons, emigrated to New York, and shortly afterward to San Francisco. The first "American" scientist to win a Nobel Prize, a Polish-Prussian Jew, had arrived in the country, a toddler of three, a half century before the prize itself would come into existence.

In 1856, the Michelson family moved to Murphy's, a remote mining town in Calaveras County, about halfway between San Francisco and Lake Tahoe. His father opened a dry goods store, but the family did not stay. Moving culturally ever further from their German Jewish roots, Michelson's family finally settled in a fledgling town in Nevada. The new "city," then little more than a campground on the slopes of Mount Davidson, was settled in 1859. According to legend, a drunken miner smashed a bottle of whiskey on a rock to christen the settlement. So was born what would soon be one of the largest cities in the Old West: Virginia City. But the honor to the state is only via transitivity. The miner, James "Old Virginny" Finney, had named the town after himself. The gold and silver in Mount Davidson quickly turned

Finney's town into one of the first industrial cities of the West, comparable in size to San Francisco, and, like it, flush with guns, gambling, and, of course, saloons. One of Albert's younger sisters later wrote a novel, *The Madigans,* about life there. His younger brother Charles, who became a New Deal ghostwriter for President Franklin Roosevelt, also wrote about it, in his autobiography, *The Ghost Talks.* But young Albert was to spend little time with his family after the move. Instead, showing intellectual promise, he was left with relatives in San Francisco to attend Lincoln Grammar School, and later, Boy's High School, where he boarded with the headmaster.

In 1869, young Michelson entered a competition to enroll in the U.S. Naval Academy across the country in Annapolis, Maryland. He didn't make the cut. It turned out to be a test of perseverance as much as knowledge: the sixteen-year-old hopped on the transcontinental railroad, completed only a few months earlier, and headed out to Washington to see President U. S. Grant. Meanwhile, his Nevada congressman wrote to Grant on Albert's behalf. His message: young Albert is a pet among Virginia City Jews, and if Grant could help him, it would help firm up the Jewish vote. Michelson eventually got to see President Grant. We have no record of how the meeting went. In popular culture Grant's reputation is not unlike that of Virginia City: whiskey plays an important role. Except for a short period in his life, that is an inaccurate characterization. What is true, but not often mentioned, is that at West Point, Grant had excelled in mathematics. Whether Grant's motive was a soft spot for a young math talent, or the president throwing a token to his Jewish constituency, what Grant did was extraordinary: he gave Michelson a special appointment to the academy, requiring them, for that year, to increase their strict quota on new cadets. In the long run, the

Michelson-Morley experiment may have been Grant's most important legacy.

Michelson became the school champion in boxing, and his rough-and-tumble, Wild West background became a part of his identity at the academy. Academically, Michelson finished ninth in a class of twenty-nine. But his overall rank does nothing to illuminate the real dynamics of his career: he ranked first in optics and acoustics; in seamanship, he came in twenty-fifth; and in history, dead last. Michelson's talents and interests were crystal clear. The Naval Academy's opinion of Michelson's focus was also clear. Its superintendent, John L. Worden (who had in 1862 commanded the *Monitor* in battle against the *Merrimac*), told Michelson, "If you'd give less attention to scientific things and more to your naval gunnery, there might come a time when you would know enough to be of some use to your country." Despite its apparent emphasis on shooting rather than science, the physics course at Annapolis at the time was one of the best in the country. Michelson's physics textbook there was a translation of an 1860 text by a French author, Adolphe Ganot. In it, Ganot describes a substance believed to pervade the entire universe: ". . . there is a subtle, imponderable and eminently elastic fluid called the ether distributed throughout the universe; it pervades the mass of all bodies, the densest and most opaque, as well as the lightest and most transparent."

Ganot goes on to attribute to the ether a fundamental role in most of the phenomena experimentally studied in his time—light, heat, and electricity: ". . . A motion of a particular kind communicated to the ether can give rise to the phenomenon of heat; a motion of the same kind, but greater frequency produces light; and it may be that a motion different in form or in character is the cause of electricity."

The modern concept of the ether was invented by Christian

Huygens in 1678. The term was Aristotle's name for the fifth element, the stuff the heavens were made of. According to Huygens's vision, God had made space like a big aquarium, our planet like a floating toy you drop in to amuse the fish. Only the ether, unlike water, flows not only around us but also through us. The concept had attraction to anyone who, like Aristotle, was uncomfortable with the idea of "nothingness," or vacuum, in space. Huygens adapted Aristotle's ether in an attempt to explain Danish astronomer Olaf Rømer's discovery that light from one of Jupiter's moons took time to reach the earth rather than arriving instantaneously. That fact, and the fact that light seems to move at a speed independent of its source, were evidence that light consisted of waves that traveled through space much as sound moves through air. But sound waves, like water waves, or the waves on a jump rope, were considered to be really just ordered motion of a medium, like air, or water, or rope. If space were empty, it was thought, a wave could not travel through it. As Poincaré wrote in 1900, "One knows where our belief in the ether stems from. When light is on its way to us from a far star . . . it is no longer on the star and not yet on the earth. It is necessary that it is somewhere, sustained, so to say, by some material support."

Like most new theories, Huygens's ether had the good, but also the bad and the ugly. In Huygens's theory, the bad and ugly is his teeny little assumption that the entire universe and everything in it is permeated with this extremely rarefied and hitherto unobserved gas. It caused Huygens to sweep a lot under the rug, for it is one thing to postulate an omnipresent fluid in the universe, another to reconcile it with known physical laws. Huygens's theory was not accepted in his lifetime, rejected in favor of Newton's view of light as particles.

In 1801, an experiment was performed that altered the prevailing view. It also provided the most important new tool for the study of light in the nineteenth century. The setup seemed innocent, a variation on experiments that had been done for centuries, the shining of light through a slit. But the English physicist Thomas Young shone *two* beams of light from a single source though two separate slits, then looked at their overlap on a screen. What he found was an alternating pattern of light and dark: an interference pattern. Interference has an easy explanation in terms of waves. Overlapping waves can add in some regions and cancel each other in other areas, like the battle crests and troughs we see in colliding water ripples. With the wave theory of light, the ether theory saw a renaissance.

It's not that the objections to Huygens's theory had dissolved over the centuries. Instead, it became a battle of repugnancy. In the far corner was light, as wave motion without a medium. Like a water wave in the absence of water, this contender was hard to cheer for. In the near corner was light, as wave motion in a medium that was everywhere present but nowhere detectable. Like water you insist is everywhere present, but nowhere to be seen, this combatant was also an undesirable. To be (but have no effect), or not to be? That was the question. To the layman, making a distinction may seem like splitting hairs. To scientists of the day, there was a clear winner: the ether. Anything was better than "not to be." The fact that physicists didn't know what ether consisted of was "of no consequence," opined E. S. Fischer in his *Elements of Natural Philosophy* (1827).

One who did not feel the nature of the ether was irrelevant was the French physicist Augustin-Jean Fresnel. In 1821, he published a mathematical treatise on light. Waves can oscillate in two ways that are fundamentally different—either along their direction of motion, like sound waves, or oscillations along a Slinky, or at right angles to it, like the waves in a rope. Fresnel showed that light waves were most probably the latter. But this kind of wave requires the medium to possess a certain elastic quality—roughly speaking, a certain amount of body. Because of this, Fresnel asserted, the ether is not a gas permeating the entire universe, it is a *solid* permeating the entire universe. What had merely been bad and ugly was now almost inconceivable. Yet for the rest of the century, it remained the accepted view.

23. The Stuff of Space

RYING TO understand what space was made of led to perhaps the greatest scientific breakthough ever. It was an intense struggle carried out for the most part by scientists who didn't know where they were going or where they were when they got there. Like space itself, their path was full of bends and curves.

● ■ ▲

The stage was set in 1865, when a 5-foot-4 Scottish physicist published a paper called "A Dynamical Theory of the Electromagnetic Field." He followed it in 1873 by a book, *A Treatise on Electricity and Magnetism*. The author's birth name was James Clerk, but in order to qualify for an inheritance from a dead uncle, his father later added the name Maxwell. As it turned out, with a little money and this unusual clause, the uncle had immortalized his name, if only among physicists and historians of science.

Maxwell's theory of electromagnetism ranks with mechanics, relativity, and quantum theory as one of the cornerstones of modern physics. You won't find his earnest, bearded face adorning coffee mugs. Neither New York nor Hollywood culture vultures consider him appealing. Yet his life is celebrated among those who, in high school and early college, labored to understand the varied and complex phenomena of electricity, magnetism, and light, and then, after learning vector calculus, suddenly discovered that the phenomena are all contained in a few innocent lines almost akin to what Alexei would call "number sentences." A Pasadena shop near the Caltech campus once carried a shirt with a quote paraphrased from God's lead story in the book of Genesis. It read: "God said, 'let there be [four equations were listed].' And there was light." The equations were Maxwell's equations. A handful of letters and odd symbols, and apart from gravity, these equations explained every force then known to science.

Radio, television, radar, and communication satellites are just some consequences of this knowledge. A quantum version of Maxwell's theory is the most accurately and extensively tested quantum field theory in existence; it served as the model for today's "standard model" of elementary particles, the smallest particles of matter we know of. A careful analysis of Maxwell's theory implies both special relativity and the absence of any kind of ether.

None of this was apparent at the time.

Physics students today are presented with Maxwell's theory as a concise set of differential equations determining two vector functions from which all optical and electromagnetic phenomena in the vacuum can in principle be derived. It is a beautiful theoretical development. But studying it in the texts is about as close to how its meaning was really discovered as a Lamaze class is to giving birth; there is something about the absence of intense pain and screaming that gives the two experiences a somewhat different feel. A long time ago, a young graduate student (me) turned in a homework paper in which he solved a complex electromagnetic radiation problem in two ways in order to get a better feel for the apparent magic of the more powerful solution. The elegant solution, employing modern tensor techniques, took less than a page of calculations. The "brute force" method required eighteen pages of mathematics to obtain the same answer. (The professor teaching the class subtracted points for making him sort through it all.) The latter technique was closer to Maxwell's original theory, but still not quite as awkward. Maxwell's 1865 theory consisted of a set of twenty differential equations in twenty unknowns.

One can hardly fault Maxwell for not utilizing simplifying notations not yet invented or widely used. On the other hand, Maxwell's theory was not just complicated and complicated-looking; it was poorly explained. Apparently the same meticulous nature that allowed Maxwell to absorb and unify the vast knowledge of the day, and to juggle in his head so complex a theory, interfered with his ability to explain it. As Hendrick Antoon Lorentz, one of those most responsible for interpreting and simplifying the theory, later wrote, "It is not always easy to comprehend Maxwell's ideas. One feels a lack of unity in his book due to the fact that it records faithfully his gradual transition from old to new ideas." In the less kind words of Paul Ehrentest, it was "a kind of intellectual jungle." Maxwell had given his colleagues a core dump, not a pedagogical explanation. Despite the obtuse presentation of his theory, Maxwell was the greatest master of electromagnetic phenomena the world had yet encountered. Given Maxwell's insight, what was his stand on what space was supposed to be made of? The ether, or not the ether? He published an article on it in the ninth edition of the *Encyclopaedia Britannica* in 1878:

> Whatever difficulties we may have in forming a consistent idea of the constitution of the ether, there can be no doubt that the interplanetary and interstellar spaces are not empty, but are occupied by a material substance or body, which is certainly the largest, and probably the most uniform body of which we have knowledge.

Even the great Maxwell could not let go.

To his great credit, Maxwell did not simply wave his hands as many did, and dismiss the ether as an unobservable necessity. He discovered the first, and the essential, observable consequence: If light waves travel with a constant speed relative to the ether, and if the earth moves in an elliptic orbit through the ether, then the speed at which light emanating from space approaches the earth will vary, depending on where the earth is in its orbit. After all, the earth is traveling in a different direction in January than it is in July, when it is at the opposite end of its orbit. On April 23, 1864, Maxwell attempted to perform an experiment to determine how fast the earth is moving through the ether.

He submitted an article on his effort entitled "Experiment to determine whether the Motion of the Earth influences the Refraction of Light" to the *Proceedings of the Royal Society.* Sadly, his article was never published because its editor, G. G. Stokes, convinced Maxwell that his approach was unsound. It wasn't, at least in principle. Maxwell did not live to see the question of the ether settled, but in 1879, while in agonizing pain from the stomach cancer that would shortly take his life, he sent a letter on the subject to a friend. His letter would eventually lead to experimental proof that the ether does not exist.

Maxwell's letter was published posthumously in *Nature,* where Michelson saw it. It gave him an idea for an experiment. To understand Michelson's setup, imagine Nicolai, Alexei, and their dad playing ball at the park. They are positioned with Dad at the vertex of a right angle, Nicolai to the north, and Alexei at an equal distance to the west, along the vertical and horizontal arms.

Imagine that all three of them are running north at the same speed. Let's assume Dad is 10 yards from either boy, and that they all run at 10 miles per hour. Dad is chasing Nicolai, who has run away with the ball, and Alexei keeps pace with his dad, at a constant distance along a parallel path. Dad glances at his watch and yells, "Time to go home!" As soon as the kids hear him, they yell back, "No!" Here's the question: Will Dad hear one of his sons before the other?

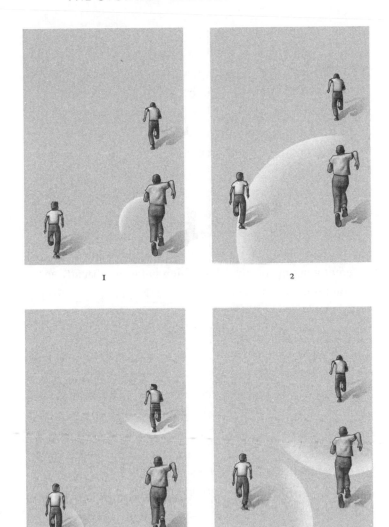

I   2

3   4

A MOVING EXCHANGE

The answer is "Yes." Regardless of how fast any of the speakers run, their yells will travel through the stationary air at the same speed, call it c. But Nicolai is running away from his dad's yell, so the yell must travel farther than their separa-

167

tion of 10 yards to reach him—it must travel 10 yards plus the distance Nicolai runs in the time it takes the yell to reach him. Nicolai's return yell, on the other hand, needn't travel the entire 10 yards to reach his dad, because his dad is running toward it. It travels only 10 yards minus the distance Dad runs in the time it takes the yell to reach him. Another way of saying this is that Dad's yell approaches Nicolai at a speed of c minus 10 mph, and Nicolai's yell approaches his dad at a speed of c plus 10 mph. Alexei, on the other hand, is running neither toward nor away from his dad, so their yells will approach their targets at the speed of just plain c.

Given this analysis, it seems clear that the round trips take different amounts of time; but which is faster, a steady speed of c on both legs of the round trip, or the slower $c - 10$ followed by the faster $c + 10$?

Alexei and Nicolai know the answer from a story sometimes read to them (as they try to avoid going to sleep). Its moral is, slow and steady wins the race. To see this, let's suppose, temporarily, that the speed of sound, c, equals 10.00001 (decimal notation for "10 plus a wee bit") mph. In that case, Alexei and his dad's yells are exchanged at the speed of 10.00001 mph, which translates to about 2 seconds each way. Nicolai's return yell will approach his dad much faster than that, at a speed of $c + 10 = 20.00001$ mph, or in about 1 second. But first Dad's yell must reach Nicolai. How long does that take? Dad's yell closes the gap at a speed of merely $c - 10 = 10.00001 - 10 = .00001$ mph. At that rate, it will take over three weeks. Alexei wins. Of course, the speed of sound is really about 675 mph, or roughly 330 yards per second. Though that makes it a photo-finish, the outcome of the race is the same.

Replacing sound with light and the air with the ether, the above experiment becomes an embodiment of Maxwell's idea. Dad and the boys won't have to run, either, because the earth is already hurtling through space, orbiting the sun at a speed of about 18.5 miles per second. (The earth is also revolving on its axis, but at a much lower speed.) There is one subtle point: that the earth is moving around the sun at a given speed does not imply it is moving through the ether at that speed. Yet it seemed to imply that the earth must be moving through the ether at *some* speed, a speed one would expect to vary with the season, as the earth's direction in space changes with its orbit. In fact, our experiment with Dad and the kids should allow us to measure the earth's speed through the ether, since knowing by how much Alexei wins allows us to solve for the speed c. This is essentially the experiment Michelson performed. The experiment was simple, except in that laboratory we call the real world.

Light moves fast, even compared to the speed of earth in orbit, about 10,000 times as fast. It is a conveniently round number for the theory, but a nightmare for the experiment. The mathematics of the situation tells us that at this tiny speed the difference in time between Alexei and Nicolai's exchanges with Dad is only one millionth of a percent. That means that if Dad, Alexei, and Nicolai were a light-year apart, the signals from the kids would still return within about a third of a second. Is the method practical? It seemed not.

Fortunately for Michelson, a Frenchman named Armand-Hippolyte-Louis Fizeau had been left a fortune by his father, a doctor, and used his time and money to pursue his interest in optics. Fizeau was interested in particular in designing a terrestrial apparatus to measure the speed of light, an undertaking Galileo had once envisioned. But Galileo did not have the benefits of the industrial revolution, and the advances in precision tooling that came in the mid-nineteenth century. To accomplish his goal, Fizeau succeeded in building an apparatus in which a beam of light traveled uninterrupted over a path of 5 miles. Five miles is a long trip on a slow bus, but it goes by rather quickly at 186,000 miles per second. Still, in 1849, Fizeau's measurement came within 5 percent of what we now know is the speed of light. In 1851, he carried out a series of experiments to test a theory that the ether is dragged along the earth's surface. Proposed in 1818 by Fresnel, this theory proved important, for it meant it was possible that points on the earth's surface might have a small or zero velocity with respect to the ether, after all. Fizeau's 1851 apparatus was complex and impressive, and it contained an important innovation—a "beam splitter," made from a lightly silvered mirror which allows a beam of light to be split into two beams that take different paths and are later recombined. In Michelson's setup, a thin beam of light from a tiny light source impinged upon such a mirror, half of it continuing through, the other half reflecting at 90 degrees. Dad's role as the vertex was played by this half-silvered mirror. Alexei and Nicolai were replaced by normal mirrors, which simply reflected back toward the vertex the light that it had sent them.

Michelson used a constant and tiny light source to create a narrow beam to send into the splitter. Since light acts as a wave, if on recombination one beam had returned faster than the other, the oscillations of the two beams would no longer be in phase, i.e., in step with each other. This would produce interference, which can be translated into a time difference to determine our speed through the ether as before. (If not for the need to use this interference effect as a means of measurement, the experiment could have been accomplished merely by shining a light between two points in different orientations and comparing the times of transit.)

Michelson couldn't actually hope to have the two arms of his apparatus equal to within a wavelength, or even to measure their length to that accuracy. Moreover, he had no way of knowing what angle his apparatus would actually make with the ether velocity. Michelson cleverly solved these problems by rotating his apparatus through 90 degrees and measuring the shifts in the fringes as the two beams "exchanged roles" rather than measuring the fringes themselves.

Michelson had not had to travel far to develop as a boxer, but as a scientist, the situation was quite different. In 1880, he received permission from the navy to journey across the Atlantic to further his studies. A fellowship of this sort was common at the time, an attempt by the U.S. government to compliment its military brawn with a sprinkling of military intelligence. Not yet thirty, Michelson developed his brilliant idea for the interferometer during stays in Berlin and Paris.

Michelson's proposed apparatus had to be built to state-of-the-art precision: a change of a thousandth of a millimeter in the distance of one arm relative to the other could ruin his measurements. If the temperature of one arm were only a hundredth of a degree higher than that of the other, Michelson's experiment would fail. Before he began, Michelson would cover the arms with paper to prevent temperature changing drafts, and surround his instrument with melting ice to keep its perimeter at a uniform 0 degrees Celcius. In the end, his apparatus was so sensitive that it could detect the disturbance made by a stomp on the pavement 100 yards from the lab.

Such an apparatus is costly. Michelson wanted the brass framework built by the famous German instrument makers Schmidt & Haensch, but he couldn't afford such an extravagance. Fortunately, a fellow American had achieved fame and fortune a few years earlier through his invention of the "talking telegraph," a little device we today call the telephone. By 1880 its inventor, Alexander Graham Bell, was working on a new invention—the videophone. Bell had contracted Schmidt & Haensch for research instruments, and had an account there. It was on this credit account that Michelson's apparatus was built.

Michelson performed his experiment in Potsdam, Germany, in April 1881. He found virtually no time difference between the two paths though space. What did this mean? Michelson's purpose had not been to disprove or even to test the ether hypothesis: it was to *measure our speed* through the ether. When he found nothing, he did not conclude that the ether did not exist; he merely concluded that somehow we weren't moving through it. How could the earth not be moving through the ether? One answer was the ether drag theory provided by Fresnel and seemingly confirmed, though not with such precision, by Fizeau. In any case, if Michelson did not see his work as a challenge to the existence of the ether, neither did anyone else. Sir William Thomson (Lord Kelvin), on a visit to the United States in 1884, put it bluntly: ". . . the luminiferous ether is . . . the only substance we are confident of in dynamics. One thing we are sure of, and that is the reality and substantiality of the luminiferous ether." The bottom line was that Maxwell's electromagnetic theory requires waves, and waves require a medium. Most physicists ignored Michelson's experiment altogether. He later wrote, "I have repeatedly tried to interest my scientific friends in this experiment without avail. . . . I was discouraged at the slight attention it received."

● ■ ▲

One of those who took Michelson's experiment very seriously was the Dutch physicist Lorentz. In 1886, he questioned Michelson's theoretical analysis, pointing out a problem actually first mentioned by the French physicist André Potier in 1882. Michelson's analysis, like ours above, does contain a subtle error. In our discussion, we assumed that Dad's yell to Alexei travels horizontally (in our setup) from Dad's position at the time of the yell to Alexei's at the time Alexei hears it. But by the time the yell reaches Alexei, everyone has moved upward a little. This means that

Dad's yell has to travel farther than the 10 horizontal yards we assumed it did. This extra bit of distance accounts for a little extra time and reduces the amount by which their interchange beats Nicolai and Dad's. In the new analysis, the shifting of the interference fringes was only half that originally expected by Michelson. Lorentz argued that if the correct analysis were used, Michelson's experiment would involve enough experimental error to invalidate Michelson's conclusion.

Michelson returned to the United States, to a professorship at Case School in Cleveland. Before long, Lorentz and also Lord Rayleigh were calling for a refinement and repetition of the experiment. Michelson began to work on it with a colleague from across the fence at neighboring Western Reserve College, Edward Williams Morley. Then, in 1885, Michelson suffered a nervous breakdown and left school for New York. Morley kept working, not expecting Michelson back, but he did return by the end of the semester. In Cleveland at noon on July 8, 1887, and again on the 9th, 11th and 12th, Michelson and Morley performed the definitive experiment that has become a part of every physics student's curriculum. Reaction to the refined experiment was as tepid as before. The negative result now seen as revolutionary seemed to many nothing more than a failure to find the desired effect—a measurement of our speed through the ether. Though Michelson and Morley had planned further measurements, for instance at different seasons, i.e., points along the earth's orbit, they too, lost interest.

Like the discovery of curved space, the Michelson-Morley experiment did not produce an explosion in the history of ideas. It was more like the lighting of a fuse. The first wisp of smoke from that fuse appeared in 1889, when, after the experiment seemed long forgotten, a short letter appeared in a new American journal, *Science*. The letter began like this:

I have read with much interest Messrs. Michelson and Morley's wonderfully delicate experiment attempting to decide the important question as to how far the ether is carried along by the earth. Their result seems opposed to other experiments showing that the ether in the air can be carried along only to an inappreciable extent. I would suggest that almost the only hypothesis that can reconcile this opposition is that the length of material bodies changes, according as they are moving through the ether or across it, by an amount depending on the square of the ratio of their velocities to that of light. . . .

What could this mean? The length of material bodies *changes?* The space we live in changes matter? The letter finished with just two more, long sentences. Written by an Irish physicist, George Francis FitzGerald, it described a form of one of the fundamental concepts of the theory that would finally explain Michelson and Morley: relativity.

At around the same time, Lorentz, who was still pondering the Michelson measurements, came to the same conclusion. Only Lorentz, the leading theoretical physicist of the 1890s, tried to construct an explanation of the contraction of bodies based on the way molecular forces are transmitted through the ether. (By now, in an effort to save the idea, the ether was no longer assumed to be unaffected by physical forces.) Without a physical explanation for the contraction, it was an ad hoc appendage, like Ptolemy's epicycles. Yet attempts to formulate one failed, particularly because the forces Lorentz was forced to postulate were hard to reconcile with Newtonian mechanics.

● ■ ▲

By 1904, a year before Einstein's first paper on relativity, Lorentz and others made several curious discoveries, but did not appreciate their implications. Lorentz's new theory differentiated between two types of time, "local time" and "universal time" (but considered universal time to be somehow the favored measure). Lorentz had also realized that an electron's movement through the ether must affect the value of its mass, an effect confirmed experimentally by the physicist Walter Kaufman. Poincaré questioned whether the speed of light might be a speed limit for the universe, a law seemingly implied by the contraction theories. He also speculated about the subjectivity of space and time, writing: "There is no absolute time; to say two durations are equal is an assertion which by itself has no meaning . . . we have not even direct intuition of the simultaneity of two events occurring in different places. . . ." The dividing line between temporal things and the timeless space in which they existed in was breaking down. What kind of geometry would emerge then?

It took Albert Einstein to formulate a simple theory that explained the observed behavior of light traveling through space. Space and time were forever joined and their uncle geometry grew very eccentric indeed.

24. Probationary Technical Expert, Third Class

HEN NAPOLEON, in 1805, rode past Gauss's home in Göttingen, he was returning from a decisive victory at Ulm. Napoleon spared Göttingen out of an appreciation for Gauss, but the site of his victory would soon become equally hallowed as the birthplace of arguably the greatest physicist in history: Albert Einstein. It was 1879, the year Maxwell died.

Unlike Gauss, Einstein was no boy genius. He was late to begin speaking—some say three. As a young child, he was generally quiet and withdrawn. He was tutored at home until one day he had a tantrum and threw a chair at his teacher. In elementary school, his record was spotty. At times he did well; but other teachers thought him dull-witted, perhaps even retarded. Unfortunately, as today, rote learning was the focus of most schoolwork, and rote learning was never one of Einstein's strong suits. Quick to appreciate a child who might immediately yell "north" when asked in which direction a compass points, they had little appreciation for one who instead ponders, as Einstein did at age five, what invisible force might cause it to do so. Not that the German schools hadn't made progress since the days of Buettner and Gauss. The punishment for a wrong answer was no longer a whipping. The more modern technique was a sharp rap on the knuckles. The hidden genius behind Einstein's often-slow-in-coming responses was actually the pain-aversion strategy of a frightened child: he would always mentally check and recheck his answer before speaking.

At his parent-teacher conference, nine-year-old Albert's mom and dad might have heard something like this: Young Albert is good in math and Latin, but well below grade level in everything else. One imagines his teacher's doubts and his parents' concern. Would this fourth grader ever amount to anything? In retrospect, by age thirteen, Einstein was showing exceptional ability in mathematics. He began to study advanced math with an older friend and with an uncle. He also studied the work of Kant, especially Kant's views on time and space. Kant may have been wrong about the role of intuition in mathematical proof, but his idea of that time and space are the products of our perception interested Einstein even as a teenager. Though human psychology plays no role, the subjectivity of measurements of space and time are what give relativity its name.

By 1895, young Einstein also knew of the Michelson-Morley experiment, of Fizeau's work, and of that of Lorentz. Though at this time he accepted the ether, he had realized that, no matter how fast you move, you must never be able to catch up to a light wave. Relativity was percolating.

Einstein's outside intellectual pursuits did not reflect themselves in an easier time at school. When Albert was fifteen, his Greek teacher, apparently not the nurturing type, pronounced in class that the boy was intellectually hopeless, was wasting everyone's time, and should leave school immediately. He said it, wisely, in German rather than Greek, or Albert probably would not have understood him. Albert didn't leave immediately, but he did soon take his teacher's advice. He got a note from a family doctor that he was heading for a breakdown, and another from his math teacher saying he already knew all the math in the curriculum. He took the notes to his principal and was allowed to drop out.

At the time, Albert was living in a boardinghouse—his family had moved to Italy. Now Albert was free to follow them there. He may have been driven from school for no good reason, but he found that the dropout's life suited him. The future guru of physics and rival of Isaac Newton's spent the next six months prancing around Milan and the surrounding countryside. When asked about job plans, he said that a real job was out of the question. What he would consider was a job teaching college philosophy. Unfortunately, university philosophy departments weren't hiring a lot of high school dropouts. And even teaching high school required a college diploma. You don't have to be an Einstein to figure out the only option left was simply to have a good time.

But Albert's father, Hermann, was also an Einstein, and this Einstein wasn't about to let that happen. Recognizing his son's mathematical talent, he nagged, cajoled, and, to put it in his native Yiddish, *hocked a chainik* until Albert agreed to go back to school to study electrical engineering. Hermann himself was not an electrical engineer, but he had started a couple of electrical equipment businesses (both failed). Albert chose to apply to one of the best schools—the Eidgenoessische Technische Hochschule (ETH) in Zurich, known by many different names in English, most of which have in common the word "Polytechnic." The university was internationally famous—and also one of the few universities that did not require a Gymnasium degree. Instead, all one had to do was to pass an entrance exam. Albert took it. He flunked.

As usual, Albert did well on the math part of the exam, but, as usual, there were a few other pesky subjects included on the test. In this case French, chemistry, and biology dragged him down. Since he probably had no ambition to write French biochemistry papers, it must have seemed pointless to Albert to bar him from school on these grounds. It seemed pointless to others as well. Albert was applying to the big leagues now, and in the big leagues, his mathematical promise did not go unnoticed.

Heinrich Weber, a mathematician and physicist who was the school's professor of physics, invited Albert to audit his lectures. The principal, Albin Herzog, arranged for him to get another year's preparation at a nearby school. The next year, Gymnasium diploma in hand, Einstein was allowed into the ETH without retaking the test. Einstein rewarded Weber and the principal's faith by validating their entrance exam: as it portended, he did poorly in school. And why not? The curriculum suffered from the same defective educational philosophy as the exam. As Einstein put it: "One had to cram all this stuff into one's mind for the examinations, whether one liked it or not. This coercion had such a deterring effect on me that, after I had passed the final examination, I found the consideration of any scientific problems distasteful to me for an entire year."

Einstein muddled through by studying the notes of a friend, Marcel Grossmann, who would play a key role in Einstein's later mathematical life. Weber was not amused by Einstein's behavior and considered him arrogant. The fact that Einstein considered Weber's lectures obsolete and not worth attending probably had something to do with that. His charming manner had turned Weber from mentor to nemesis. Three days before Einstein's final exam in the summer of 1900, Weber decided to get even: he required Albert to rewrite an article he had turned in because his submission was not written on regulation paper. To those born after 1980: in precomputer days you couldn't accomplish this by simply reloading your printer and clicking a mouse. It involved tedious application of something called handwriting. This consumed much of Albert's remaining study time.

Einstein came in third out of the four students, but passed. His fellow graduates got university jobs, but Weber, giving him bad recommendations, stood in Einstein's way. Einstein did a stint of substitute teaching, and then some tutoring, and on June 23, 1902, ended up in his now-famous employment in the Swiss Patent Office. His glamorous title was Probationary Technical Expert, Third Class. While working at the Patent

Office, Einstein completed a Ph.D. at Zurich University. In later years, he recalled that his thesis was at first rejected for being too short. He added a single sentence and resubmitted it. This time, it was accepted. It is hard to tell if the story is true or was just a bad dream after a night of too much cognac, because there seems to be no evidence to back it up. Nevertheless, it captures the essence of Einstein's academic life to that point.

His "education" behind him, in 1905 Einstein's brain exploded with revolutionary ideas, enough to win him three or four Nobel prizes, were they given out by any objective criteria. It was the most productive year any scientist had had, at least since Newton's 1665–66 stay at his mother's farm. And Einstein did not have the leisure to sit and watch apples fall—he did it all while working full time at the Patent Office. His production consisted of six papers (five of them published that year). One was based on his doctoral thesis, a matter of geometry—not the geometry of space, but the geometry of matter. Einstein published his dissertation in *Annalen der Physik* under the title "A New Determination of Molecular Dimensions." In it, he presented a new theoretical method for determining the size of molecules. This work later found application in a wide variety of arenas from the motion of sand grains in cement mixes to casein micelles (particles of protein) in cow's milk. According to a study done by Abraham Pais in the 1970s, between 1961 and 1975 it was quoted more than any other scientific paper written prior to 1912, including Einstein's papers on relativity. Einstein also wrote two papers on Brownian motion in 1905. This is the irregular motion of tiny particles suspended in liquid first noticed by the Scottish botanist Robert Brown in 1827. Einstein's analysis, based on the idea that the motion is due to the random bombardment of the particles by molecules of the liquid, led to a confirmation of the new molecular theory of matter by French experimentalist Jean-Baptiste Perrin. Perrin received the Nobel Prize for this work in 1926.

In another paper written in 1905, Einstein gave an explanation of why certain metals had been observed to emit electrons when light shone upon them, an effect known as the photoelectric effect. The main issue to be explained was that, for a given metal, there existed a threshold frequency below which the effect did not occur no matter how intense a beam of light you employed. Einstein applied Max Planck's quantum idea to explain the threshold—if light consisted of particles (later dubbed photons) whose energy was dependent on frequency, then only above certain frequencies would the impinging photon have enough energy to dislodge an electron.

In this paper, Einstein brashly applied Planck's new quantum concept as if it were a universal physical law. At the time, it had been considered merely a poorly understood facet of the interaction between radiation and matter. This worried no one, for it was a field that at the time was full of question marks anyway. Certainly no one dared imagine, as Einstein did, that the quantum idea might apply to radiation, thereby contradicting the well-understood and well-tested theory of Maxwell. As with Einstein's other revolutionary work, at first few were convinced. Lorentz and even Planck himself opposed Einstein's view. Today we look at Einstein's paper as a landmark in the history of quantum theory, a step on par with Planck's discovery of the quantum itself. For this, Einstein received the 1921 Nobel Prize for Physics. But it was Einstein's other two 1905 publications for which, a century later, he is most remembered. They represented the beginning of an eleven-year odyssey that led scientists into the strange new universe of curved space which Gauss and Riemann had shown was mathematically possible.

25. A Relatively Euclidean Approach

N TWO papers in *Annalen der Physik* in 1905, "On the Electrodynamics of Moving Bodies," published September 26, and "Does the Inertia of a Body Depend on Its Energy Content?" published in November, Einstein explained his first theory of relativity, special relativity.

In his Gymnasium days, Einstein had discovered a book on Euclid. Unlike Descartes and Gauss, Einstein was a fan: "Here were assertions, as for example the intersection of the three altitudes of a triangle in one point, which—although by no means evident—could nevertheless be proved with such certainty that any doubt appeared to be out of the question. This lucidity and certainty made an indescribable impression on me." Ironically, in his later theories, non-Euclidean geometry was to play a central role. But for special relativity, Einstein took Euclid's approach. He based his reasoning for special relativity on two axioms about space:

1. It is impossible to determine, except in comparison to other bodies, whether you are at rest or in uniform motion.

Einstein's first axiom, usually called the principle of relativity, or Galilean relativity, was first postulated by Oresme. It holds true even in Newtonian theory. One day recently, Nicolai was riding through the apartment atop a plastic fire truck. Alexei, absorbed in a pre-teen horror novel, sat on a chair in our drive-through kitchen. As he whizzed by, Nicolai held out a plastic ax, thoughtfully included when we bought the truck and helmet. As he passed, Nicolai's ax knocked into Alexei's book, causing both book and ax to fall to the ground, and inspiring the usual charges and countercharges. Alexei argued that his passing brother stuck an ax at him, knocking the book out of his hand. Nicolai claimed that he was holding his ax still and Alexei moved into it. Dad, preferring not to investigate questions of judicial consequence, broke into a lecture on the science of the situation.

Newton's laws predict the same events whether Nicolai was stationary and Alexei's book moving, or Alexei was stationary and Nicolai's ax in motion. This is Einstein's first postulate—you cannot distinguish one case from the other, so each kid's point of view is equally valid. (Both got a time-out.)

2. The speed of light is independent of the speed of the source and is the same for all observers in the universe.

Like the first, Einstein's second axiom was also not revolutionary in itself. As we have seen, Maxwell's equations required the speed of light to be independent of the source, and this bothered no one because it is the normal behavior of propagating waves. The meat in Einstein's assumption is contained in the phrase "and is the same for all observers." What does this mean?

If you could tell whether you are moving, it wouldn't mean much: all observers could agree that the speed of light was its speed in approaching a "stationary" object. This is the situation within Newton's framework—absolute space, or the ether, provides a frame of reference against which motion can be measured. But if you cannot distinguish rest from uniform motion, and all observers measure the same speed of approaching light, whether or not they themselves are in relative motion, then we encounter the spitting paradox we mentioned earlier. How could a light wave be approaching both you and the spit at the same speed?

To understand how light can behave this way, we must question what is behind our reasoning. Given that we wish to take Einstein's two axioms as—well—axiomatic, we won't question them. What other assumptions have we made? We have made heavy use of the concept of simultaneity, so it seems natural to examine that. This is just what Einstein did.

Let's consider a situation akin to one Einstein himself employed in his 1916 book, *Relativity*. Einstein liked to make use of railroad car analogies, because in his experience train rides provided the starkest real-world evidence that it is impossible to tell if you are in uniform motion. Anyone who has ridden a train or subway today has probably had the experience Einstein drew on, of not being sure whether it is your car or the neighboring one (or both) that is in motion. In our example, Alexei and Nicolai are on opposite ends of a subway car. It is their first subway ride alone. Mom and Dad stand on the platform, waving, hoping the *Out of Service* signs they stuck on the windows will keep this car relatively uncrowded. Suppose Mom and Dad stand with the same separation as Alexei and Nicolai, in a way that, shortly after the train begins to move, Mom will be even with Alexei and Dad with Nicolai. This serves a purpose: they brought cameras. Mom, because this is their sons' first ride, and Dad so that he'll have a good photo for the police when the kids don't return at the appointed time. Due to the law of nature called sibling rivalry, Mom and Dad plan to snap their shots at precisely the same moment, Mom capturing Alexei's smiling face and Dad Nicolai's. The photos being simultaneous, neither son can brag that his photo was taken first. Nevertheless, the stage is set for a family feud.

The cause of the feud lies in the answer to this simple question asked by Einstein: Are two events judged to be simultaneous by the parents also judged to be simultaneous by the kids? Our first question is, What does it mean to say that the two events occurred simultaneously? If two events occur at the same place, the answer is trivial: They are simultaneous if they occur at the same time (as measured by a clock at that point). It takes real insight to understand that the answer is not so trivial if they do not occur at the same place.

Suppose light (or anything we could use to send a signal) moved at infinite speed. Then, the moment the flashes went off, both flashes would immediately reach both Alexei and Nicolai. They could then each answer the question of simultaneity the simple way—by comparing events at a point, in this case the arrival of the two flashes at their location. If they perceived one flash first, then that photo was taken first. But since light does not travel at infinite speed, this method won't work. Dad, always the scientist, has a suggestion. He sets up photo detectors along the way between him and Mom. If the photos are taken at the same time, the light from their flashes should meet midway. Nicolai, having heard Dad's idea, repeats it as his own (one of his more endearing habits). Alexei sets up photo detectors in their subway car.

The train begins to move. Mom and Dad have synchronized watches. The photos are snapped. Sure enough, the light meets halfway between them. Are Alexei and Nicolai satisfied? No, because by the time the flashes meet, their car will have moved a little, so the flashes will not meet at the halfway point in their car. The situation is depicted in the figure on the following page.

From the kids' point of view, each flash is an event that occurs at a time and place in their world, the subway car, which they justifiably perceive to be at rest. Like their parents, they see no reason the flashes should not meet halfway. So when the flashes meet nearer to Alexei, they conclude that Nicolai's photo was snapped first. Though the photos were timed by Mom and Dad to be simultaneous, that is not the way it appears in a frame of reference that is moving with respect to

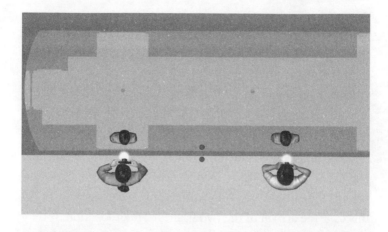

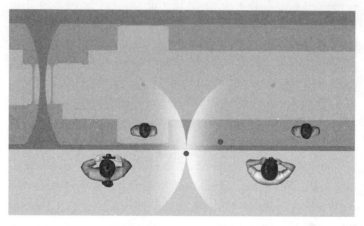

SUBWAY TIME

them. Dad is kicking himself for not having arranged things differently—so that the flashes were not simultaneous to him but would have been to the kids.

Yes, you may say, I see the point, but whom are you kidding? The kids are the ones really moving, while the parents are on the motionless platform. It may seem that way because we think of the earth as motionless, but of course it is not. Imagine an observer in outer space—with the earth hurtling

186

around the sun, and spinning to boot, it may seem petty indeed to insist that either the train or the platform is somehow more naturally considered "at rest." Or, strip away the props: imagine the kids and parents out in empty space. Now there is truly no external way to tell who is moving. The effect is the same and it is real—what appears simultaneous to the parents will not appear so to the kids, and vice versa.

With the fall of simultaneity comes the relativity of distance and time. To see this we need only note that to measure the length, we need to first mark off the endpoints of what we wish to measure, then hold a yardstick up against it. If the object is at rest with respect to us, this is trivial. But if it is moving, there is an intermediate step. We could, for instance, mark the two endpoints on, say, a stationary piece of paper as the object passes. Then, as before, we can hold up the yardstick to measure the distance between the two marks. But we must make sure that we mark off the endpoints—here comes that nasty word again—simultaneously. If we err and mark one end before the other, the latter end will have traveled some distance and we'll not get a true measurement. Unfortunately, when we make what we perceive as simultaneous measurements, a person moving with the measured object will, as we have just seen, not agree. That person will accuse us of marking one end before the other, and obtaining a faulty result. This means that objects do not have length in an absolute sense. Their length depends on the observer who is looking. This is a new kind of geometry.

It is often said that in relativity, moving objects appear contracted in their direction of motion. This means that an object, when measured by an observer who considers the object to be moving, will appear shorter than when measured by a person to whom the object appears at rest. Einstein found analogous anomalies in the behavior of time. Observers in motion with respect to each other will not agree on the length of an interval of time, or on how much time has passed. Like length, duration has no absolute meaning.

The time that an observer measures between two events at his own location—which, in his frame of reference, is a fixed point in space—is called the *proper time.* Any other observer who is moving (at a constant velocity) with respect to this observer will perceive the two events to occur with a longer time interval between them. Since we are always at rest with respect to ourselves, ignoring the effects of acceleration, our lifetime, as measured by us, will always appear shorter than it appears to others. To others our clocks will seem to run slow. But we will die, alas, on the cue of the internal timer that travels with us. In special relativity the grass *is* greener on the other person's lawn.

What does this mean for the laws of motion? In special relativity objects still follow Newton's first law: they move in a straight line unless acted upon by an external force. Observers may disagree on how long a particular segment of the line is, but not about whether it is straight. Yet this is not a "relativistic way" of phrasing the first law: in relativity space and time mix differently for different observers. The concepts of geometry must be altered to embrace time as well as space.

Instead of points in space and times of occurrence, we formalize the term *events,* that is, points in the four dimensions of space and time. Instead of paths through space, we speak of *worldlines* through space and time. Instead of distance, we have a combination of the time interval and the spatial distance between events. And instead of lines, we consider geodesics, defined now (for technical reasons) as either the shortest or longest worldline connecting two events. A typical event is this writer sitting at the particular point in space that is his desk, at a particular time. A typical worldline is the writer at his desk for many hours on end. That particular worldline has a time coordinate that varies, but space coordinates that do not. That is allowed for worldlines. The "path" he has taken in space is a boring, fixed point, but in space-time he still traces out a worldline, just as a rising elevator traces a path that is constant in east-west coordinate, but changes only in elevation. The distance in space-time between two points on that worldline is not zero, though their distance in space is zero— because the points are separated in time.

To discover how to rephrase Newton's first law in relativistic language, suppose an object is to travel from Alexei at time zero on his clock, to Nicolai at time one second on his, as objects often seem to do. What path will the object take in the absence of an external force? In the language of relativity, the two events we consider are (space = Alexei's location, time = zero) and (space = Nicolai's location, time = one). Assuming the kids are stationary with respect to each other, and that they have synchronized clocks, the object will move in a straight line at whatever constant speed it takes to get from Alexei to Nicolai in one second on their clocks. This is the worldline of a free object in special relativity.

What is the law governing this worldline? Consider what would have been different if the object hadn't moved in a straight line, that it took a detour. Having more distance to cover in the same time, it would have had to move faster to reach the target on time, i.e., to reach the event (Nicolai's location at time = one second). But as we have seen, when an object moves relative to another, its clock will seem to run slower: the object will arrive with less than one second elapsed time on its clock.

Motion in a straight line in space, and at a constant speed, forms the worldline for which an object's clock reads the *maximum* possible elapsed time between two events. Newton's first law can thus be stated, employing the new geometry, like this:

> Unless acted upon by an external force, an object always follows a worldline from one event to another in a way that the time read by its own clock (i.e., the proper time) is at a maximum.

Einstein knew his theory was a cannonball tossed into the castle that was modern physics. He idolized Newton, but he was destroying one of Newton's most basic beliefs, the existence of absolute space and time. He was also obliterating a two-century-old cornerstone of physical theory—the ether. Though his special theory of relativity had many triumphs (the explanation of the longer lifetimes of fast-moving radioactive particles, the equivalence and convertibility of energy and matter), Einstein was smart enough to know that those who had spent their lives maintaining and decorating the castle might not offer schnapps and a pat on the back to the guy who destroyed it. He braced himself for attack.

Months passed, and no attack came. Issue after issue of *Annalen der Physik* appeared, and, regarding Einstein's bombshell, the world of physics seemed to have nothing to say. Finally, Einstein received a letter from Max Planck asking for clarification on a couple points. More months passed. Was that it? You pour your soul into a revolutionary new theory of nature, and all that comes from it are a few questions from a guy in Berlin?

On April 1, 1906, Einstein was promoted to technical expert, second class, at the Patent Office. An honor by Patent Office standards, but not exactly a Nobel Prize. He began to wonder if he was, as Alexei would say, a transplant from the planet Loser. Or, in Einstein's own words, a "venerable federal ink shitter." To make matters worse, at twenty-seven, Einstein was afraid that his creative days were numbered. He might have wondered if he'd die in obscurity like Bolyai and Lobachevsky, but like almost everyone else, he'd never heard of them.

What Einstein didn't know was that the letter he had received was the tip of the Max Planck iceberg. In the winter semester of 1905–06, Planck presented a physics colloquium on Einstein's theory in Berlin. And in the summer of 1906, he sent one of his students, Max von Laue, to visit Einstein at the Patent Office. Finally, Einstein would have his chance to interact with the world of real physicists.

When Einstein entered the room where von Laue was waiting, he was too shy to introduce himself. Von Laue gave him a glance, but ignored him because von Laue did not imagine so unimpressive a man could be the author of relativity. Einstein left. A little later, he returned, but still couldn't muster the courage to approach von Laue. Finally, von Laue introduced himself. As they walked to Einstein's house, Einstein offered von Laue a cigar. Von Laue sniffed it. Cheap and awful. While they spoke, he surreptitiously dropped it into the Aare River. Von Laue was unimpressed by what he saw and sniffed, but he was very impressed by what he heard. Both von Laue, who would win a Nobel Prize in 1914 (for his discovery of X-ray diffraction), and Max Planck, who would win one in 1918, became key supporters of Einstein and relativity. Years later, recommending Einstein for a position in Prague, Planck would compare him to Copernicus.

Planck's support of relativity was ironic in view of the hard time he had accepting Einstein's earlier explanation of the photoelectric effect, a new interpretation of Planck's own quantum theory. Yet when it came to relativity, Planck was open-minded and flexible, and immediately recognized it as correct. In 1906, Planck became the first other than Einstein to publish a paper on relativity. In that paper he also became the first to apply relativity to quantum theory. And in 1907 he became the first to direct a Ph.D. dissertation on relativity.

Einstein's former teacher at ETH, Hermann Minkowski, then at Göttingen, was another to fly the banner of relativity. One of the few in the early days to make an important contribution to the theory, he gave a colloquium in which he introduced geometry and the idea of time as the fourth dimension into the theory of relativity. In a 1908 lecture, Minkowski said, "Henceforth space by itself, and time by itself, are doomed to fade away into mere shadows, and only a kind of union of the two will preserve an independent reality."

Despite support from a core of physicists, mainly in Germany, broad acceptance for special relativity was slow in coming. In July 1907, Planck wrote to Einstein that the advocates of relativity "form a modest crowd." From many, acceptance never came. Michelson, as we have seen, could not let go of the ether. Lorentz, who shared mutual respect with Einstein, couldn't quite make the break either. And Poincaré, who never understood relativity, continued to oppose it until his death in 1912.

But as the physics community slowly pondered Einstein's ideas, he began working toward a second, even greater revolution. It would be a revolution that would again make geometry the centerpiece of physics, a spot from which it had strayed since Newton's introduction of the equations of calculus. It would also be a revolution that would make Einstein's first, in comparison, seem easy to swallow.

26. Einstein's Apple

S EINSTEIN later told it, in November 1907, "I was sitting in a chair in the patent office at Bern when all of a sudden a thought occurred to me: 'if a person falls freely he will not feel his own weight.'"

Thoughts like this were not what Einstein was getting paid for. He was there to reject perpetual motion machines, analyze ideas for a better mousetrap, debunk contraptions for turning dung into diamonds. The work was occasionally interesting, and never very taxing. But the hours were not short: eight hours a day, six days a week. Still, he would work on his physics after hours. In later years it came out that he also would often bring his notes in to work on them secretly at the office, quickly stuffing them into his desk when the director approached. Herr Einstein, a schlump like the rest of us. The director was so out of touch that when, in 1909, Einstein finally resigned to take a university position, he laughed and thought Einstein was joking. Brownian motion had been explained, the photon invented, and the special theory of relativity developed, all right under his nose.

"It a person falls freely he will not feel his own weight." Einstein later called it "the happiest thought of my life." Was Einstein a sad, lonely man? Actually, his personal life was no Hollywood fairy tale. He was married, divorced, remarried, and remained negative about married life. He gave up his firstborn for adoption. His youngest child was schizophrenic and died in a psychiatric hospital. He was chased by the Nazis from his home continent and never completely at ease in his adoptive country. But the thought that so greatly pleased Einstein would have held prominence in anyone's life, had it had the same significance.

Einstein said the realization "startled" him; it was the epiphany that led to his greatest accomplishment. The falling person of Einstein's thought was Einstein's apple, the seed whose progeny was a new theory of gravitation, a new concept of cosmology, and a new approach to physical theory. Ever since 1905, Einstein had been looking for something like this, a new principle that could act as a guide in his search for a better theory of relativity. He knew his original theory was incomplete. With all its implications for the subjectivity of space and time, in the end, his special theory of relativity was only a new kinetics. It described how bodies react to specified forces; it did not specify them. Of course, special relativity was designed to mesh perfectly with Maxwell's theory, so the specification of electromagnetic forces was not a problem. Gravitational forces were a different story.

The only theory of gravity in 1905 was Newton's. Being a smart fellow, Newton designed his description of gravitational forces to dovetail perfectly with his kinetics, i.e., with his laws of motion. Since special relativity replaced Newton's laws with a new kinetics, it is not surprising that Einstein found that Newton's theory of gravity no longer fit. Newton's theory of the gravitational force is this:

> The gravitational attraction between two point masses at any given instant is proportional to each of the masses, and inversely proportional to the square of their distance at that instant.

That's all there is, really. You can translate it to mathematics to make it quantitative. You can apply calculus to graduate from "point" masses to extended ones. You can plug it into his laws of motion to produce the equations that govern how objects like the heavenly bodies move under each other's influence. Or, in what first made Gauss famous, with a lot of sweat and genius, you can solve those equations (approximately) to predict the orbit of a newly discovered asteroid, in Gauss's case, Ceres. Discovering the consequences of Newton's law of gravitation was far more complex than its simple statement, and physicists had no trouble extracting from it thousands of person-years of work.

Newton himself was unhappy with his law; he considered the instantaneous transmission of force to be a suspicious concept. In relativity, it is just plain guilty: nothing can be transmitted faster than the speed of light. And there's more. Consider the phrase "at a given instant." In relativity, as we have seen, this is a subjective judgment. If the two masses are moving relative to each other, events that to one mass would appear simultaneous would to the other seem to occur at different times. Neither, as Lorentz found, would they agree on the values of masses, or distances.

Einstein knew that for his theory to be complete, he must find a description of gravity that is consistent with special relativity. But something else also bothered Einstein. He had made a big point in special relativity of the principle that an observer should be able to consider himself at rest without having to change his theories of physics, like the principle that the speed of light is a given constant. This should apply to any observer. But in special relativity it applied only to an observer in uniform motion.

"What is this favored state of being called uniform motion?" a skeptic, or a logician, might ask with a snarl. The well-practiced answer is, the state of motion in a straight line with constant speed. True, a set of observers moving in straight lines and at constant speed *with respect to each other* makes a fine "old boys' club" whose members can all smugly agree on their uniformity. But can they debunk the rebel view of an outsider who says their motion is uniform *merely* with respect to each other, and only because, in reality, they are all changing speed or direction in unison?

Imagine a sports stadium full of fans glued to their seats by the thrill of the game. They seem like the epitome of uniform motion—that of the couch potato (uniform motion with speed zero). But now imagine another couch potato, this time an astronaut glued to her seat watching the game on a TV monitor as she reclines on a space-station BarcaLounger. To her, all those fans in the stadium are spinning madly about the earth's axis, hardly what she would call straight-line motion. What judge can adjudicate her claim that she is the one at rest and they are spinning? Or, now that the floodgates have opened, another observer's claim that both she and the stadium are moving insanely, in unison jerking this way and that?

As it happens, there is a way to tell. For the author of this book, it is simple: under uniform motion, he sits calmly and ponders how beautifully Newton's laws describe the world around him; subjected to too many accelerations, he turns green and vomits. It's an effect first observed in a Chevy in the early sixties. The effect of acceleration on the human body is of course complex, but the physics behind it is simple: acceleration makes a difference. Imagine a thought experiment with Einstein's son Hans Albert as the guinea pig. Hans Albert was five years old in 1907, an age in which highly non-uniform motion still seems perversely attractive. Imagine Hans Albert on a merry-go-round, and his dad, Dr. Einstein, on the fixed platform that encircles it.

Hans Albert holds a lollipop in his hand. He lets go. If the merry-go-round is standing still, the lollipop will merely fall to the ground. If it is spinning, the lollipop will fly away along the tangent line at the moment of its release. Young children tend to view themselves as the center of the universe. Suppose Hans Albert does this, insisting that in both cases he is at rest. In the latter case, he will not view the merry-go-round as being in motion. Instead, in his view, the world will be orbiting him. What bothered the elder Einstein was that, unlike the scenario in which Nicolai's ax collided with Alexei's book, in these two observers' descriptions the events seemed to obey different rules. To see this, let's examine how both observers analyze the situation. Einstein the father would lay down a coordinate system fixed to the earth. In this system, his position would be unchanging and Hans Albert's path would be a circle around the center of the merry-go-round. The lollipop, for a while, would travel with Hans Albert, forced by his grasp to follow in his circular path. At the moment Hans Albert let go, the lollipop would continue according to Newton's laws of motion. That means it would leave the circle and begin moving in a straight line with whatever speed and direction it had the instant Hans Albert relaxed his grasp. Neither Newton's laws nor special relativity requires any modification to describe what is happening.

Now consider little Hans Albert's point of view. He lays down a coordinate grid affixed to the merry-go-round, one on which he does not change position. The lollipop remains for a while at rest, at Hans Albert's position. But when Hans Albert opens his hand, the lollipop suddenly flies away. This is not the usual behavior of objects in either Newton's or Einstein's physics. Their laws do not appear to apply. Instead, in this frame of reference, Hans Albert might be tempted to replace Newton's first law with a statement like this:

> An object at rest tends to remain at rest, but only if you hold it tight. If you let go, things fly away from you for no apparent reason.

A rotating observer like Hans Albert, who insists on considering himself at rest, would have to change the laws of physics to describe how objects moved in his world. Altering Newton's laws of motion (i.e., kinetics) is only one way to do that. If Hans Albert cared to "save" Newton's laws, he could do this instead: keep Newton's laws, but define a mysterious "force" that acts on everything in the universe, shoving it away from the center of the merry-go-round. Since, except for being repulsive instead of attractive, this sounds a bit like gravity, let's call the force *schmavity*.

Newton knew that accelerated motion of a frame of reference makes objects move as if mysterious forces like schmavity act upon them. Such apparent forces were known as *fictitious forces* because they did not arise from a physical source such as a charge, and could be eliminated if one looked at the situation from a different reference frame, one in uniform motion (called an *inertial* frame). The absence of fictitious forces, in Newton's theory, provided the true criterion for uniform motion. If no fictitious forces appeared, you were in uniform motion. If they did, you were accelerating. This explanation bothered many scientists, especially Einstein. Okay, in this sense uniform motion seemed definable, physically. But in the absence of the fixed framework of absolute space, does it really make any more sense to single out accelerating frames of reference than it did to single out a particular frame as being at rest?

Imagine a test object in a space devoid of all matter and energy. How do you distinguish between linear and circular motion when there is nothing to measure motion against? Newton had answered this question with his belief in absolute space: even completely empty space came endowed with a fixed framework that would define motion. God wasn't the "batteries not included" type—the universe came equipped, not only with Euclid but also Descartes. A popular alternative theory of the day was a proposal by Austrian physicist Ernst Mach: the center of mass of all the matter in the universe defines a point relative to which all motion is judged. Thus, roughly speaking, motion that is uniform relative to the distant stars is true inertial motion. But Einstein had his own ideas.

With special relativity, Einstein had succeeded in obliterating the distinction between rest and uniform motion (at a non-zero speed); he had inertial observers on equal footing. He now sought to widen his theory to embrace all observers, including those accelerating with respect to inertial frames. If he succeeded, his new theory would require no fictitious forces to account for "non-uniform motion," nor would the statement of the physical laws of motion need to change. The couch potatoes in the stadium, the astronaut on the moon, Hans Albert on the merry-go-round, Albert himself on the fixed platform, would each be able to employ his theory without thought to what might be the true inertial frame of reference. The philosophical motivation was there; all Einstein lacked was the theory. How to approach it? He needed a guiding principle.

The realization Einstein had as a result of his "happiest thought" gave him just what he needed. "If a person falls freely he will not feel his own weight." That was the first sign post, and the compass on the long road toward a new theory. Stated more broadly, the statement became the equivalence principle, or Einstein's third axiom:

> It is impossible to distinguish, except in comparison to other bodies, whether a body is undergoing uniform acceleration or is at rest in a uniform gravitational field.

In other words, gravity is a fictitious force. Like schmavity, it can be considered to be a mere artifact of our chosen frame of reference, and can be eliminated by choosing a different one. This principle applies to a uniform gravitational field and this is its simplest form, the form in which Einstein first thought of it. The work of Gauss and Riemann allowed Einstein to apply it to any gravitational field by considering a non-uniform field to be a patchwork of infinitesimal (i.e., really small) uniform fields patched together, but he didn't assert that until five years later, in 1912. It was then, also, that he coined the term *equivalence principle.*

Let's see what Einstein meant in the original case of the uniform field. In visualizing uniformly moving frames of reference, Newton used ships to ponder frames of reference in the same way Einstein used trains, and sometimes elevators. Newton might have viewed gravity differently if he had had the elevator, but that mode of transport didn't start to catch on until after 1852, the year Elisha Graves Otis solved a little engineering problem: how to keep passengers from plunging to their death when the cable snapped. Einstein used pre-Otis elevators in his thought experiments regarding general relativity. Suppose, riding in an elevator, you suddenly feel weightless. The equivalence principle is merely the embodiment of this intuitive observation: in this circumstance you cannot determine whether the cable has been cut, or whether gravity has simply disappeared (though the latter might be considered wishful thinking). If an environment is allowed to fall freely in a uniform gravitational field, the laws of physics are the same as they are for a gravity-free environment. Let go of your coffee and it will just float there, whether you are in outer space or in the process of falling to your death from the ninety-first floor.

Now suppose you step into an elevator at the ground floor of an office building. The doors slide shut. You close your eyes. You open them. You feel your usual weight. What causes you to feel this downward force? It might be earth's gravity, or the earth may have been suddenly obliterated by aliens and your elevator hijacked upward, accelerating, adding to its speed by 32 feet per second each second. It's not a speculation you'd want to share on a job interview, but according to the equivalence principle, the effect of both scenarios is identical. Let go of your coffee and it will splash the same either way.

That objects in a freely falling elevator appear to float, or objects in an accelerating elevator in gravity-free space appear to fall is of course predicted by Newton's laws. There is no new physics per se in these scenarios. But as usual, Einstein was relentless in interrogating the situation until it confessed its hidden secrets. The secrets he heard from this one were strange ones—the presence of gravity must affect the passage of time and the shape of space.

To find the effect on time, Einstein applied an analysis within the elevator that was in the same spirit as the one he applied to the subway. He traced the perceptions of various observers exchanging and timing light signals. Einstein planned to use special relativity to describe the physics, but he encountered a problem. Since these observers were accelerating, special relativity did not apply. So he made an assumption that later became one of the cornerstones of his final theory: that within a small enough space and a short enough time, and for a small enough acceleration, special relativity approximately applies. In this way, Einstein could apply special relativity, and the equivalence principle, to infinitesimal regions, even in a non-uniform field.

Imagine a long rocket ship with Alexei at the top and Nicolai at the bottom. They have identical clocks. Alexei starts to flash a light at each tick of his clock. For simplicity, let's suppose that by Alexei and Nicolai's measurements, the spaceship is one light-second long. (Meaning a flash of light would take one second to travel from Alexei to Nicolai.) What does Nicolai observe?

Since Alexei creates a flash every second and each flash

travels the same one light-second to reach Nicolai, after one second Nicolai will observe a flash every second. Now suppose the rocket blasts off with constant acceleration. What changes? The next flash will arrive sooner than expected because Nicolai will have traveled toward the flash. Say it arrives 0.1 second early. According to the equivalence principle, Nicolai and Alexei may deny that any motion has occurred and instead attribute the "pull" they feel to a gravitational field. But if they deny acceleration and attribute the force to a gravitational field, then they would deny, too, that Nicolai has traveled upward to meet the flash. Instead, they would conclude from the fact that the signal arrives 0.1 second early that the application of the gravitational field has caused Alexei's clock to speed up, causing him to release the flash 0.1 second early.

If, as the equivalence principle specifies, either interpretation must be allowed, we are forced to conclude that a clock that is located higher in a gravitational field will run faster. Due to the earth's gravitation field, time for Alexei in the top bunk moves just a little quicker than it does for Nicolai on the bottom bunk. Very very little. Even with the sun's vastly greater gravitational field, time on earth, 93 million miles above it, runs only about two parts per million faster than time at the sun's surface. At that rate, a being on the sun earns only about an extra minute a year. Hardly worth the trade-off in climate. This warpage of time affects the frequency of light, which is the number of oscillations of the light wave per *second*. It's not a big effect, but it is one that Einstein predicted (called the gravitational redshift). Because of this, if your favorite station were AM 1070 (i.e., 1070 kHz), broadcasting from atop the 110-story World Trade Center, the frequency you would want to tune into on the ground would be AM 1070.00000000003. Hi-fi buffs take note.

Einstein first made the argument that the passage of time is altered by gravity in 1907. We know from special relativity that space and time are intertwined. How long did it take for the probationary technical expert to realize that the presence of gravity thus also alters the shape of space? Five years—a good point to remember next time you overlook something you think should have been obvious. As Einstein said, "If we knew what it was we were doing, it would not be called research, would it?"

Einstein made the space-warping connection in Prague, in the summer of 1912. It was his sixth year pondering his gestating general theory of relativity. Once again, it came as an epiphany. He wrote, "Because of the Lorentz contraction in a reference frame that rotates relative to an inertial frame, the laws that govern rigid bodies do not correspond to the rules of Euclidean geometry. Thus Euclidean geometry must be abandoned. . . ." In English, "when you don't move in a straight line, Euclidean geometry is distorted."

Imagine Hans Albert, then ten, again on the rotating merry-go-round. And suppose that to his dad on the "stationary" platform, the merry-go-round appears to have the shape of a perfect circle. What does special relativity say about space in this situation? (As before, this analysis is not strictly speaking rigorous, because it involves the application of special relativity to non-uniform motion.) Consider, at each instant of time, drawing two perpendicular axes at Hans Albert's instantaneous position. One axis points radially (out from the center of the merry-go-round). This is the direction of the force Hans Albert feels at that moment. Hans Albert is not moving at all in this direction: his distance from the center of the merry-go-round is unchanging. The other axis is a tangent to the merry-go-round. At any given instant it points along the direction of Hans Albert's motion. It is always perpendicular to the force he feels.

Now suppose his dad tosses Hans Albert a tiny horizontal

square, with one side lined up along the radius of the turning platform. He asks Hans Albert to observe it and report back its shape. What will Hans Albert say? What to his dad was a square to him appears as a rectangle. This is the effect of Lorentz contraction. Since Hans Albert is always moving tangentially, and never radially, the two sides of the square parallel to the tangent are contracted; the sides parallel to the radius are not. If Hans Albert measured the circumference and diameter of the merry-go-round in terms of these lengths respectively, he would find that the ratio is not equal to π. Hans Albert's space is curved. His father concludes that Euclidean geometry must be abandoned. His only question was, abandoned in favor of what?

27. From Inspiration to Perspiration

T IS EASY to abandon, harder to construct. What Einstein needed if he was to construct a new physics was a new geometry that described the warpage of space. Fortunately, Riemann (and a couple of later followers) had worked this out. Unfortunately, Einstein hadn't heard of Riemann—hardly anyone had. But Einstein *had* heard of Gauss.

Einstein remembered a course on infinitesimal geometry he had taken as a student. It had covered Gauss's theory of surfaces. Einstein approached his friend, Marcel Grossmann, to whom, in 1905, Einstein had dedicated his doctoral thesis. Grossmann was by then a mathematician in Zurich who happened to specialize in geometry. On seeing him, Einstein exclaimed, "Grossmann, you must help me or I'll go crazy."

Einstein explained his need. As he searched the literature, Grossman discovered the work of Riemann and others on differential geometry. It was arcane. It was complex. It wasn't pretty. Grossmann reported back that, yes, such mathematics did exist, but it was "a terrible mess which physicists should not be involved with." But Einstein did want to mess with it. He had found the tools to formulate his theory. He also discovered that Grossmann was right.

In October 1912, Einstein wrote a friend and fellow physicist, Arnold Sommerfeld, "in all my life I have labored not nearly as hard, and I have become imbued with a great respect for mathematics. . . . compared with this problem, the original theory [special relativity] is child's play."

The quest took another three years, two of them in close collaboration with Grossmann. The student whose notes got Einstein through college again became his tutor. Planck, upon hearing what Einstein was up to, told him, "As an older friend I must advise you against it for in the first place you will not succeed; and even if you succeed, no one will believe you." But by 1915 Einstein was back in Berlin, attracted there by Planck himself. Grossmann wrote only a handful of research papers after that, and in less than a decade was severely ill with multiple sclerosis. Einstein, having learned what he needed, completed his theory without him. On November 25, 1915, he presented a paper entitled "The Field Equations of Gravitation" to the Prussian Academy of Sciences. In it, he announced, "Finally the general theory of relativity is closed as a logical structure."

How does general relativity describe the nature of space? It shows how the matter and energy of the universe affect the distance between points. Viewed as a set, a space is simply a collection of elements, its points. The structure of space we call geometry arises from the relation between points we call distance. The added structure is like the difference between a phone book, listing homes, and a map defining their spatial relationship. During his time spent surveying Germany, Gauss discovered that by defining the distances between pairs of points, you determine the geometry of the space; Riemann developed the details Einstein needed to phrase physics in these terms.

It all boils down to the dispute between our old friends, Pythagoras and Nonpythagoras. Recall that, in a Euclidean world, we can measure the distance between any two points by employing the Pythagorean theorem. We simply lay down a rectangular coordinate grid. Let's call the two coordinate axes the East/West axis and the North/South axis. According to the Pythagorean theorem, the square of the distance between the points is equal to the sum of the squares of their East/West separation and their North/South separation.

As Noneuclid found, in curved space such as that of the earth's surface, this is no longer true. Instead, the Pythagorean theorem must be replaced by a new formula, the Nonpythagorean theorem. In the Nonpythaogorean formula for distance, the North/South term and East/West term are not necessarily counted equally. Also, a new term is possible, the product of the East/West and North/South separation. Mathematically, this reads: (distance)^2 = g_{11} × (East/West separation)^2 plus g_{22} × (North/South separation)^2 plus g_{12} × (East/West separation) × (North/South separation). The numbers represented by the g-factors are called the *metric* of the space (the g-factors are called the *components* of the metric). Since the metric defines the distance between any two points, geometrically, the metric completely characterizes the space. For the Euclidean plane, with rectangular coordinates, the metric's components are simply g_{11} = g_{22} = 1, and g_{12} = 0. In that case, Nonpythagoras' formula is just the usual Pythagorean theorem. In other types of space the components are not that simple, and their values can vary depending on your location. In general relativity, these ideas are generalized to three spatial dimensions, and, as they were in special relativity, to include time as a fourth dimension (in four dimensions the metric has ten independent components).

Einstein's 1915 paper announced this: an equation relating the distribution of matter in space (and time) to the metric of four-dimensional space-time. Since metric determines geometry, Einstein's equations define the shape of space-time. In Einstein's theory, the effect of mass is not to exert a gravitational force, but to change the shape of space-time.

Though space and time are intertwined, if we restrict ourselves to certain circumstances, namely, low speeds and weak gravity, then space and time can be viewed as approximately separate. In this realm, it is acceptable to speak of ·

space alone, and of the curvature of space. According to Einstein's theory, the curvature of a region of space (averaged over all directions) is determined by the mass within the region.

As we have seen, curvature is reflected in the relationship of the area of a circle to its radius, or the volume of a sphere to its radius. Einstein's equations reflect this: given a spherical region of space with matter uniformly distributed within it, the measured radius of the sphere will be less than the radius you'd expect (given its volume) by a factor proportional to the amount of mass within it. The proportionality constant is extremely small: for each gram of mass, the radius is off by only 2.5×10^{-29} centimeters, i.e., .00000000000000000000000000025 centimeter. For the earth, assuming a uniform density, that makes the excess radius 1.5 millimeters. For the sun, it is half a kilometer.

The manifestations of space-time curvature on earth are minute and have only recently had practical application (Global Positioning Satellites, for instance, require general relativistic corrections to stay in sync). For years Einstein did not think the bending of light by gravity was measurable at all. Then he thought to look skyward. The test is simple in principle: Look up where and when the next full eclipse of the sun will occur; measure the position of a star which will appear next to the sun during the eclipse (hence the need for the eclipse: if the sun were not blocked, spotting this star would be hopeless); also find its position from other data from, say six months earlier, when its light could travel to your eyes without grazing our own star. During the eclipse, check to see whether its image appears where it "should" be, or whether it is "off" a little.

A little, in this case, was really a little: only 1¾ arcseconds, or .00049 degree. Newton himself could have discovered the same effect, though his theory would have predicted a different deflection. By 1915, Einstein had discovered his field equations and made his best prediction. The first real test of general relativity, then, was not whether light bent, but how much it bent. Einstein was confident.

28. Blue Hair Triumphs

TWO BRITISH expeditions were sent to make observations during the eclipse of May 29, 1919. Arthur Stanley Eddington led the one that was to be successful, to Sobral in Brazil. Before leaving, Eddington wrote: "The present eclipse expeditions may for the first time demonstrate the weight of light [i.e., its attraction by gravity—the "Newtonian" analysis]; or they may confirm Einstein's weird theory of non-Euclidean space; or they may lead to a result of yet more far-reaching consequences—no deflection." It took months to analyze the data. Finally, on November 6, the result was announced at a joint meeting of the Royal Society and the Royal Astronomical Society. The *New York Times,* which until then had never once mentioned Einstein's name, sensed that this was news fit to print. It may have misjudged the importance of the news, however—it sent its golf specialist, Henry Crouch, to cover the announcement. Crouch didn't even attend the meeting, but he did talk to Eddington.

The next day, *The Times* (London) headline read: "REVOLUTION IN SCIENCE," with smaller headlines, "New Theory of the Universe" and "Newtonian Ideas Overthrown." *The New York Times* report appeared three days later, with the headline: "EINSTEIN THEORY TRIUMPHS." *The New York Times* article praised Einstein, but it also questioned whether the effect might not have been an optical illusion, or whether Einstein might have stolen his idea from H. G. Wells's novel *The Time Machine.* They got Einstein's age wrong, calling him "about fifty" when he was forty. But though the *Times* got his age wrong, they spelled his name right. Around the world, Einstein became an instant celebrity, to many an almost supernatural genius. One starry-eyed schoolgirl wrote to ask him if he really existed. Within a year there were more than a hundred books written on relativity. Lecture halls around the world overflowed with people anxious to hear popular expositions of the theory. *Scientific American* carried an offer of $5,000 for the best 3,000-word explanation. (Einstein remarked that he was the only one amongst his friends who did not enter the contest.)

But if many in the public idolized Einstein, some of his colleagues attacked. Michelson, then head of the physics department at the University of Chicago, accepted the Eddington observation, but refused to endorse the theory. Michelson's counterpart in the department of astronomy said, "The Einstein theory is a fallacy. The theory that the 'ether' does not exist and that gravity is not a force but a property of space, can only be described as a crazy vagary, a disgrace to our age." Nikola Tesla also ridiculed Einstein, but Tesla, it turned out, was also afraid of round objects.

One day over dinner recently Alexei expressed his latest artistic desire: to dye his hair blue. This is the twenty-first century and kids have been dyeing their hair blue for at least a couple decades now. Not many, though, at age nine. The next Monday, Alexei became the first in his school to have hair that matched the color of his ink. And Nicolai, his four-year-old echo, came away with a shock of lime green hair up front.

Reaction at school was about as expected. A few kids demonstrated intellectual depth and insight and pronounced the look cool (mostly Alexei's friends). A lot of kids couldn't accept the break with tradition and called him names like "blueberry." His teacher stared for a moment, but had no comment.

Physics is a lot like the fourth grade. To physicists of the early twentieth century, non-Euclidean space was a fringe area of study. A curiosity, perhaps, but like blue hair, not very relevant to the mainstream. Then Einstein came along and proposed that blue hair become fashion. The resistance, in Einstein's case, lasted a few decades, but it gradually faded as the old generation died out and the new accepted whatever made the most sense, which was definitely not a solid permeating all space called the ether.

The last hurrah of the anti-relativists was in Germany, the country of its earliest supporters. In Germany, the anti-Semites had a field day. Nobel Prize winner (1905) Philipp Lenard and Nobel laureate (1919) Johannes Stark supported those who considered relativity a plot for the Jews to take over the world. In 1933, Lenard wrote: "The most important example of the dangerous influence of Jewish circles on the study of nature has been provided by Einstein with his mathematically botched-up theories. . . ." In 1931, a booklet entitled *A Hundred Authors Against Einstein* was published in Germany. Reflecting the mathematical sophistication of the group, it actually listed 120 opponents. Few were well-known physicists.

Einstein's old supporters Planck and von Laue did not jump ship, causing Stark, in a speech to celebrate the opening of an institute named for Lenard, to turn on them:

> . . . unfortunately, his [Einstein's] friends and supporters still have the opportunity to continue their work in his spirit. His principal promoter, Planck, still heads the Kaiser Wilhelm Society, his interpreter and friend, Mr. von Laue, is still permitted to play the role of advisor for physics in the Academy of Sciences in Berlin, and the theoretical formalist, Heisenberg, the spirit of Einstein's spirit, is even to be distinguished with a university appointment.

Heisenberg rewarded Nazi kindness by heading their effort to develop the atomic bomb. Fortunately, he didn't know his relativity *that* well—and they were beaten by brilliant Americans like Italian Enrico Fermi, Hungarian Edward Teller, and German Victor Weisskopf. Einstein stayed above the fray, generally answering neither the serious challengers nor the crackpots.

Einstein was in Pasadena, in the midst of a planned two-month stay at the California Institute of Technology, when German president von Hindenburg appointed Hitler chancellor. Storm troopers soon raided Einstein's Berlin apartment and his summer home. On April 1, 1933, the Nazis seized his property and offered a reward for his capture as an enemy of the state. He was traveling in Europe by then, and decided to seek asylum in the United States, at the new Institute for Advanced Study in Princeton. Apparently the deciding factor in Princeton's favor (over Caltech) was an offer to accept his assistant, Walther Mayer, as well. Einstein arrived in New York on October 7, 1933.

Einstein spent his later years attempting to create a unified theory of all forces. To accomplish this, he would have had to reconcile general relativity with Maxwell's theory of electromagnetism with the theories of the strong and weak nuclear forces, and, most important of all, with quantum mechanics. Few physicists believed in his program of unification. Famous Austrian-American physicist Wolfgang Pauli dismissed it, saying "What God has torn asunder, let no one put together." Einstein himself said, "I am generally regarded as a sort of petrified object, rendered blind and deaf by the years. I find this role not too distasteful, as it corresponds very well with my temperament." As we'll soon see, Einstein was on the right track, but was many decades ahead of his time.

In 1955, Einstein was diagnosed with an aneurysm of the aorta in the abdominal area. It had ruptured, and was causing him much pain and blood loss. The chief of surgery at New York Hospital examined him in Princeton and suggested surgery was possible, but Einstein replied, "I do not believe in artificially prolonging life." Hans Albert, then a distinguished professor of civil engineering at the University of California, had flown in from Berkeley, and tried to change his father's mind. But Einstein died early in the morning the next day, at 1:15 a.m. on April 18, 1955. He was seventy-six. Hans Albert died of a heart attack eighteen years later, in 1973.

Looking past the resistance and hate he had to endure, and the awe and hero-worship he inspired, Einstein's contribution to geometry is perhaps best summed up by his own prosaic description. Of his revolutionary work he wrote, "When a blind beetle crawls over the surface of the globe, he doesn't realize the track he has covered is curved. I was lucky enough to have spotted it."

V

THE STORY

OF WITTEN

In twenty-first-century physics, the nature of space determines the forces of nature. Physicists flirt with extra dimensions, and the idea that, on a fundamental level, space and time may not even exist.

29. The Weird Revolution

S THERE a relationship between the nature of space and the laws governing what exists in space? Einstein showed that the presence of matter affects geometry by warping space (and time). It sure seemed radical at the time. But in today's theories, the nature of space and matter are intertwined at a level far more profound than Einstein imagined. Yes, matter may bend space a teeny bit here and, if it truly concentrates, a larger bit there. But, in the new physics, space gets more than ample revenge on matter. According to these theories, the most basic properties of space—such as the number of dimensions—determine the laws of nature and the properties of the matter and energy that make up our universe. Space, the container of the universe, becomes space, the arbiter of what may be.

According to string theory, there exist extra dimensions of space, so small that any wiggle room we have in them isn't observable in present-day experiments (though, indirectly, it may soon be). Though they may be tiny, they, and their topology—i.e., properties related to whether they are shaped, say, like a plane, or a sphere, or a pretzel, or a donut—determine what exists within them (like you and I). Twist those tiny donut dimensions into a pretzel and—poof!—electrons (and thus humans) could be banished from existence. And there's more: string theory, though still poorly understood, has evolved into another theory, M-theory, of which we know even less, but which seems to be leading us to this conclusion: space and time do not actually exist, but are only approximations of something more complex.

Depending on your personality, you may have a tendency

217

at this point either to laugh or to scream derisive remarks about academics wasting hard-earned tax dollars. As we'll see, for many years most physicists themselves had these same reactions. Some still do. But among those working in elementary particle theory today, string theory and M-theory, though still not rigorous, are de rigueur. And whether or not they, or some later derivative, prove to be some sort of "final theory," they have already changed both mathematics and physics.

With the advent of string theory, physics has veered back toward its partner, mathematics, that abstract discipline concerned, since Hilbert, with rules and not reality. String theory and M-theory are driven, so far, not by the tradition of new physical insight or experimental data, which are lacking, but by discoveries of their own mathematical structure. It isn't to toast the divining of new particles that the tequila is poured, it is to cheer the discovery that the theory describes the existing ones. Aware that such discoveries are an inversion of the usual course of science, physicists have coined for them the new scientific term *postdiction*. In a strange contortion of the scientific method, the theory itself has become the subject of the (mental) experiments; the experimentalists are the theoreticians. It is no accident that Edward Witten, today the theory's leading proponent, has won not a Nobel Prize but a Fields Medal, its mathematical equivalent. For just as geometry and matter reflect on each other, so, now, must the studies of each. Witten goes even farther, saying that string theory should ultimately be a new branch of geometry.

This is not unlike prior revolutions reforming not only the idea of space but also the way in which research on space is approached. The story of this revolution, though, is unlike the stories of prior revolutions in one important aspect: we are still in the midst of it, and no one really knows how it will turn out.

30. Ten Things I Hate About Your Theory

HE YEAR WAS 1981. John Schwarz heard a familiar voice from down the corridor. "Hey, Schwarz, how many dimensions are you in today?" It was Feynman, himself not yet "discovered," then a cult figure only in the rarified world of physics. Feynman thought string theory was nuts. Schwarz was okay with that. He was used to not being taken seriously.

One day that year a graduate student had introduced Schwarz to a new young faculty member named Mlodinow. After Schwarz left, the student shook his head. "He's a lecturer, not a real professor. Been here nine years and still doesn't have tenure." A chuckle. "Works on this crazy theory in twenty-six dimensions." Actually, the student was wrong about the last part. It had started as a theory of twenty-six dimensions, but by then was down to ten. Still, it seemed a few too many.

The theory had been plagued by other "embarrassments" through the years, a physicist's way of saying that it had implied predictions that seemed to have nothing to do with reality. Negative probabilities. Particles that have imaginary mass and travel faster than light. Through it all, at great cost to his career, Schwarz stuck with the theory.

There is a movie Alexei likes to watch about a group of high school kids, *Ten Things I Hate About You.* At the end of the film, the heroine stands in front of her class and reads a poem of the ten things she hates about her boyfriend, but it is really a poem about how much she loves him. It's easy to picture John Schwarz reciting that poem, loving the theory, sticking with it, despite or sometimes *because of* its endearing little faults.

Schwarz saw something in string theory that few others saw, an essential mathematical beauty that he felt could not have been an accident. That the theory was very difficult to develop didn't daunt him. He was trying to solve a problem that stumped Einstein and everyone else since Einstein— reconciling quantum theory with relativity. The solution couldn't be an easy one.

Unlike relativity, the first broad-ranging quantum theory didn't come for decades after Planck's discovery of the quantization of energy levels. That changed in the years 1925–27, due to the efforts of the Austrian Erwin Schrödinger and the German Werner Heisenberg. Each independently discovered—perhaps "invented" is a better word—elegant theories that explained how to replace Newton's laws of motion with other equations that embodied the quantum principles that had been inferred over the preceding decades. The two new theories were dubbed, respectively, *wave mechanics* and *matrix mechanics.* Like special relativity, the consequences of quantum theory were directly evident only in a realm removed from everyday life, in this case not the very fast but the very tiny. At first, not only was the relation of the two theories to relativity unclear but so was their relation to each other. Mathematically, they appeared as different as their discoverers.

Picture Heisenberg, a good German, suit and tie perfectly adjusted, desk in fine order. Soon to become what has been described in ways varying from "*merely* nationalistic" to "moderately pro-Nazi," he would lead the German atomic bomb effort. Derided by others after the war, he used the yes-but-my-heart-wasn't-really-in-it defense. Heisenberg conjured his theory relying heavily on experimental data, and collaborating with fellow physicist Max Born and future storm trooper Pascual Jordan. Together they created a theory encompassing the ad hoc physical rules and patterns physicists had noted over two decades. It was a process physicist Murray Gell-Mann describes this way, "They pieced it together [from experimental data]. They had all these sum rules. One time while Born was on vacation, they used them to reinvent matrix multiplication. They didn't know what it was. When Born came back, he must have said, 'but, gentlemen, this is matrix theory.'" Their physics had led them to a mathematical structure that worked.

Picture Schrödinger this way—the Don Juan of physics. He once wrote, "It has never happened that a woman has slept with me and did not wish, as a consequence, to live with me all her life." That makes this a good place to note that it was Heisenberg, not Schrödinger, who came up with the uncertainty principle.

Schrödinger's approach to quantum theory relied more on mathematical reasoning and less on experimental data than Heisenberg's. Imagine Schrödinger looking earnest, with a hint of a smile and a mess of hair reminiscent of Einstein's. He scrawls thoughtfully in a notebook not unlike one a schoolchild would carry. Make a noise and, without concern for any sort of etiquette, he sticks a pearl in each ear to safeguard against distraction. But silence is not all he needs to nourish his creativity. His wave mechanics would come, not during an extended stay in a remote monastic retreat, but in what Princeton mathematician Hermann Weyl called "a late erotic outburst in his life."

Schrödinger first wrote down his wave equation during a tryst at a ski resort while his wife was away in Zurich. It is said that the mystery woman's company kept him stimulated and insanely productive for an entire year. The collaboration being of a kind that is not usually credited, his papers on the subject listed no co-authors. This particular collaborator's name seems lost forever.

Though Schrödinger had the better working conditions, his

wave mechanics and Heisenberg's matrix mechanics were soon proved to be equivalent by the English physicist Paul Dirac. The single theory they represented was given the neutral name *quantum mechanics*. Dirac also extended quantum mechanics to include the principles of special relativity (and shared in the Nobel prizes given for quantum mechanics in 1932 and 1933). Dirac did not do this for general relativity. There is good reason: it cannot be done.

Einstein, a parent of both theories, saw very clearly the conflict between them. Though general relativity revised much of Newton's view of the universe, it maintained one of Newton's "classical" tenets: determinacy. Given the proper information about a system, be it your body or the entire universe, Newton's paradigm meant that you could in principle calculate the events of the future. According to quantum mechanics, this is not true.

It was the one thing Einstein hated about quantum mechanics. He hated it enough that he condemned the theory. He spent the last thirty years of his life searching for a way to generalize his general theory of relativity to include all the forces of nature, and, he hoped, in the process to explain the clash between relativity and quantum theory. He failed. Now, some thirty years after Einstein's death, John Schwarz felt he had the answer.

31. The Necessary Uncertainty of Being

HE SOURCE of the indeterminacy in quantum mechanics is the uncertainty principle. According to this principle, some of the characteristics of systems that are quantified in the Newtonian description of motion cannot be described with unbounded precision.

Alexei was excited recently by an old joke he had heard. A nun, a priest, and a rabbi were playing golf. Whenever he missed a key shot, the rabbi had a habit of shouting, "God dammit, I missed!" By the seventeenth hole, the priest was rather annoyed. The rabbi promised to restrain himself, but, on missing his next putt, he again shouted, "God dammit, I missed!" At this the priest warned, "If you curse again, God will surely strike you dead." On the eighteenth hole the rabbi blew it again, and again he cursed. At that the sky darkened, the winds began to blow, and a blinding bolt of lightning struck down from the sky. When the smoke cleared, the horrified priest and shocked rabbi were left staring at the smoldering remains of the nun, burnt to a crisp. Just then, from the sky above came the thundering words, "God dammit, I missed!"

Alexei says the joke is funny because it disses God—his way of saying that it presents a picture of a flawed deity making human blunders. The concept of an imperfect God, or Nature, is what bothered many physicists about quantum mechanics. Can't God specify position exactly?

This limit to the determinism of nature inspired Einstein's famous quote, "The theory [quantum mechanics] yields much, but it hardly brings us close to the secrets of the Ancient One. In any case, I am convinced that He does not play dice." If the joke had been around in his day—and it is very old—Einstein might have muttered, "The Ancient One can land a bolt of lightning wherever and whenever he wants."

With the possible exception of Schrödinger's relations with the opposite sex, there is uncertainty in everything we encounter in life. So one might wonder, why does a principle stating this obvious fact deserve such a lofty name? The uncertainty of Heisenberg's principle is an odd kind of uncertainty. It is the difference between classical and quantum theory, between the limitations of people and those of, well, God.

Give a kid a quiz: True or false—At McDonald's, a "quarter-pounder" hamburger weighs exactly a quarter pound? The cynics among them might answer "false," employing reasoning along the lines that a company that sells forty million hamburgers a day could save a lot of meat by skimming a hundredth of a pound off each burger. But we are not talking about *systematic* error: that every quarter-pounder weighs exactly .24 pounds is equally impossible. The point is that every McDonald's burger weighs a slightly *different* amount.

The difference is not just a matter of ketchup. Measuring finely enough, you would find that each hamburger has a different thickness, a unique shape, an individuality—on the microscopic scale. Like people, no two hamburgers are alike. To how many decimal places do you have to measure the burgers to distinguish between them all by weight? Since they sell over a billion each year, which is 10^9, it is at least 9. Not much chance they'll be changing its name to a .250000000 pounder.

Just as every hamburger is different, so is every measurement. Your actions in performing the measurement, the mechanical and physical state of the scale, the currents of surrounding air, the earth's local seismic activity, the temperature, the humidity, the barometric pressure, scores of minute factors are each a little different each time you repeat a measurement. Demarcate finely enough, and these will ensure that repeated measurements never agree.

That is *not* the uncertainty principle.

Where the quantum uncertainty principle goes farther is in this: it decrees that certain traits form *complementary pairs,* pairs which possess a certain *limitation:* the more precisely you measure one trait, the less precisely you will be able to measure the other. According to quantum theory, the value of these complementary quantities beyond their limiting precision is *undetermined,* not simply beyond the scope of our current instruments.

Over the years physicists have tried to argue that this is a limitation of our theory, not nature. They have suggested that lurking somewhere are "hidden variables" that are determined, but that we don't know how to measure. As it turns out, one kind of measurement we *can* make is one to rule out such hidden variables. In 1964, the American physicist John Bell explained how it could be done. In 1982, the experiment was performed: it showed that the hidden variable proposal isn't right. The limitation really is one imposed by the laws of physics.

The mathematics of the uncertainty principle say this: the product of the uncertainty of two complementary members of the pair must equal a number called *Planck's constant.*

Position is bound in one of the complementary pairs of the uncertainty principle. Its mate, momentum, is, apart from a factor of mass, the object's velocity. The marriage certificate details a limitation on the partners: the error bounds on one must increase inversely as those on the other sharpen. It is a limitation with no exception, a marriage most Catholic, with neither infidelity nor divorce. Multiply the error bound in position and that in momentum, and the number that results can be no smaller than the number of Herr Planck.

Planck's constant is a tiny, tiny, tiny number. Otherwise,

we would have noticed quantum effects much sooner (if, in such a world, we would have existed at all). Here, the adjective "tiny" means precisely "on the order of a billionth." Planck's constant is roughly a billionth of a billionth of a billionth, or 10^{-27}, of something, in this case, of a unit called *erg-grams*. Of course, the value of Planck's constant depends on the units employed. An erg-gram is a unit whose size is of the magnitude we might encounter in daily life. Imagine a 1-gram Ping-Pong ball sitting still upon a table. To most of us, sitting still means velocity zero. An experimental physicist knows that measurements without error bars have little meaning. Instead of "the ball is sitting still," her papers would use language more like "the ball is not moving faster than one centimeter per second." In classical physics, the matter would then be settled. In quantum mechanics, even this rather unimpressive precision imposes a price: it sets a limit on the precision to which the Ping-Pong ball's position can be determined.

The error bound of 1 centimeter per second leads to a limiting precision that, like Planck's constant, is tiny, tiny, tiny. Doing the math tells us that we may pinpoint the ball to within an error of 10^{-27} cm. Since the limit is not very limiting, a familiar question arises. Who cares? Until the end of the nineteenth century, no one did, or, more correctly, no one noticed. But now let's swap things like Ping-Pong balls for things like electrons. This is the swap that physicists made around the end of that century.

Remember the phrase "apart from a factor of mass" that was so cavalierly included in the definition of momentum? It might not have seemed like much at the time, but it is the reason quantum effects are noticeable in atoms and not in Ping-Pong balls.

The mass of the Ping-Pong ball was 1 gram. The mass of the electron is 10^{-27} grams. Unlike the Ping-Pong ball, an error in velocity of 1 cm/sec for the electron thus translates to error bars in momentum of 10^{-27} gm-cm/sec—due to the electron's mass factor, the velocity measurement that seemed sloppy translates to a momentum measurement that is very precise. That doesn't bode well for your ability to measure the electron's location.

If, as for the Ping-Pong ball, you determined the speed of an electron to plus or minus a cm/sec, the electron's location could not be pinpointed to better than plus or minus 1 centimeter. This limiting precision is not tiny, tiny, tiny—it is quite noticeable. It would make the game of Ping-Pong a rather sloppy one, but this is exactly the situation on the atomic scale. For electrons in atoms, simply determining that they are somewhere within the 10^{-8} centimeters we think of as the atomic confines, we are forced upon an uncertainty in velocity of 10^{+8} centimeters per second, an uncertainty about equal to the velocity itself.

Quantum mechanics, as Heisenberg and Schrödinger formulated it, was very successful in describing the phenomena of atomic physics, and also much of the nuclear physics known at the time. But when you apply the uncertainty principle to gravity, as described by Einstein's theory, you are driven toward some rather bizarre conclusions about the geometry of space.

32. Clash of the Titans

NE REASON that Einstein had little support in his search for the unified field theory was that the clash between general relativity and quantum mechanics is apparent only when you consider regions of space so small that, even today, we have no hope of observing them directly. But Euclid said that space is made of points, and geometry ought to apply to as small a region as we can imagine. If the theories clash there, there must be something wrong with one or both theories— or with Euclid.

The realm in which the problems arise is often described as the ultramicroscopic. For the quantitative among us, that means 10^{-33} centimeters, called the *Planck length*. For the visual among us, it means if you were to expand the Planck length to the diameter of a human egg cell, a typical marble would balloon to the size of the observable universe. The Planck length is *really* small. Yet, compared to a point, its size is generous beyond measure.

One night, after working on this chapter, the clash between Einstein and Heisenberg acted itself out in a dream. It all began with Nicolai, as Einstein, walking in and showing me some theories he had scrawled in crayon on his pre-K activity book . . .

Nicolai as Einstein: Dad, I've discovered general relativity! When matter is around, space is curved, but in empty space the gravitational field is zero, and space is flat. In fact, in any region that is small enough, space is approximately flat.

(I'm about to say, "What a beautiful theory, can I hang it on the wall?" when Alexei enters.)

Alexei as Heisenberg: Sorrrrry. The gravitational field, like any field, is subject to the uncertainty principle.

Nicolai as Einstein: So?

Alexei as Heisenberg: So in empty space, while the field might be zero on the average, it is really fluctuating in space and time. And in *really tiny* regions the fluctuations are *humongous.*

Nicolai as Einstein (whining): But if the gravitational field is fluctuating, so is the curvature of space, because my equations show that the curvature is related to the value of the field . . .

Alexei as Heisenberg (taunting): Ha-ha! That means the space in tiny regions cannot be considered flat. . . . In fact, when you look closer than the scale of the Planck length, tiny virtual black holes form. . . . It isn't very pretty . . .

Nicolai as Einstein: I said I want tiny regions of space to be flat!

Alexei as Heisenberg: But they aren't!

Nicolai as Einstein: They are!

Alexei as Heisenberg: Aren't.

Nicolai as Einstein: Are.

. . . In the dream this went on until I woke up palpitating. (I took it as a sign that I was not meant to sleep until I finished the chapter.)

Applying both the uncertainty principle and general relativity to tiny regions of space leads to basic contradictions with the theory of relativity itself. Who is right, Heisenberg or Einstein? If Einstein is right, then quantum theory is wrong. But quantum theory doesn't seem to be wrong. Experiment and theory agree to better than a part in a million. Cornell physicist Toichiro Kinoshita, one of the leaders in quantum electrodynamics, calls it "the best tested theory on earth, perhaps even in the universe, depending on how many aliens there are."

If quantum theory is right, relativity must be wrong. It has had its triumphs, too. But there is one difference. The triumphs of general relativity involve observations of macroscopic phenomena. Light passing near the sun or clocks flying around the earth. General relativity on the small scale of elementary particles has not been tested. Their masses are far too small for the effects of gravity to be measured. For this reason, physicists prefer to question the validity of relativity, specifically Einstein's assumptions about the approximate flatness of tiny regions of space. Perhaps Einstein's theory, for the realm of the ultramicroscopic, must be revised.

If Planck is truly the winner in the Planck-Einstein debate, and the metric of the ultramicroscopic fluctuates wildly, then another, deeper question is raised. What is the structure of space on the scale of the ultramicroscopic? The key to the answer appears to be an idea that Feynman and others had a hard time swallowing, a source of ribbing for Schwarz, but in his mind not a flaw, merely a trait of his beloved theory. In the realm of the ultramicroscopic there appear to be other dimensions, curled up on themselves, so miniscule that, like the quantum in 1899, they have been hitherto undetected. They are a key ingredient in the remedy of general relativity. They were also an idea considered and then abandoned decades ago by the creator of relativity himself.

33. A Message in a Kaluza-Klein Bottle

HE DAY before Einstein died, he asked to be brought his latest calculations on unified field theory. He had attempted for thirty years, and failed, to alter his general theory of relativity in order to include in it a description of electromagnetic forces. One of the most promising approaches came to Einstein one day in 1919, near the beginning of his quest, as he was opening his mail. The idea came, not from his own mind, but in a letter from an impoverished mathematician named Theodor Kaluza.

What Einstein read was a proposal for how one could unite electrical forces with gravity. The theory had one little quirk. Einstein wrote back, "The idea of achieving [a unified theory] by means of a five-dimensional cylinder world never dawned on me. . . ." A five-dimensional cylinder? Why would that idea dawn on anyone? No one knows how Kaluza got his idea, but Einstein wrote back, "I liked your idea enormously." In retrospect, Kaluza was ahead of his time, and a bit stingy with the dimensions.

As we have seen, general relativity described how matter affected space through the metric, whose components—the g-factors—tell you how to measure the distance between neighboring points based on their coordinate differences. The number of g-factors depends on the number of dimensions of space. For instance, in three dimensions there are 6. In flat space, distance equals (difference in x)^2 plus (difference in y)^2 plus (difference in z)^2, so g_{xx}, g_{yy}, and g_{zz} each equal 1, and the factors corresponding to cross-terms, g_{xy}, g_{xz}, and g_{yz}, each equal zero—those terms are not present. In the four-dimensional non-Euclidean space of general relativity, there are ten independent g-factors (taking into account equalities such as $g_{xy} = g_{yx}$), all described by Einstein's equations. Kaluza started by realizing that if you employed five dimensions, there would be still other g-factors corresponding to the extra dimension.

Then Kaluza asked this question: If you formally extend Einstein's field equations to five dimensions, what equations do you get for the extra g-factors? The answer was astounding: You get Maxwell's equations for the electromagnetic field! From the fifth dimension, electromagnetism suddenly crops up in a theory of gravity. Einstein wrote, "The formal unity of your theory is startling."

Of course, interpreting the metric of the extra dimension as the physical electromagnetic field takes some theoretical work. And what of that little quirk, the extra dimension? Kaluza asserted that it was finite in length, in fact, that it was so tiny that we wouldn't even notice if we jiggled within it. Not only that, Kaluza claimed, but the new dimension has a new topology, that of a circle instead of a line, i.e., it closes back upon itself, or "curls up" (and thus, unlike a finite line, has no ends). Imagine Fifth Avenue without breadth, simply a line. Cross streets, in Kaluza's new dimension, will be circles sprouting from Fifth Avenue. Of course, cross streets come at one-block intervals, but the extra dimension is present at all points along the street. So adding the new dimension to a line doesn't turn it into a line that sprouts circles, it turns it into a cylinder, like a garden hose. A very thin garden hose.

Kaluza's point was that gravity and electromagnetism are really components of the same thing, but only look different because what we observe is averaged over undetectable motion in the tiny fourth spatial dimension. Einstein had second thoughts about Kaluza's theory, but then changed his mind again and helped Kaluza publish it in 1921.

In 1926, Oskar Klein, an assistant professor at the University of Michigan, independently invented the same theory, with a few improvements. One was that he realized that the theory only leads to the correct equations of particle motion if a particle has certain values of momentum in the mysterious fifth dimension. These "allowed" values were all multiples of a certain minimum momentum. If you assume, as Kaluza did, that the fifth dimension closed in on itself, you could then use quantum theory to calculate from the minimum momentum what the "length" of the curled fifth dimension might be. If it came out to an observable, macroscopic size, the theory would be in trouble since we have never observed this new dimension. But it came out to 10^{-30} centimeters. No trouble there. It would be hidden all right.

The Kaluza-Klein theory was a hint at something, a formal connection between theories, but not a structure that immediately gave anything new. In the ensuing few years physicists looked for new predictions that might come from the theory, much in the way Klein had reasoned about the size of the new dimension. They developed new arguments that seemed to imply they could use the theory to predict the ratio of the electron's mass to its charge. But this prediction was way off. Between that difficulty and the bizarre prediction of a fifth dimension, physicists lost interest. Einstein last considered it in 1938.

Kaluza, who died the year before Einstein, never got much further. But he did benefit in one great way from his fledgling theory. When he wrote to Einstein, he was thirty-four and had been supporting his family for ten years as a *privat docent* (something like an adjunct professor) in Königsberg. His salary is best described in terms of the mathematics he loved: for each semester he received 5 times x times y German marks (or, to be technical, gold marks), where x was the number of students in his class and y the numbers of hours he lectured each week. For a class of ten meeting five hours a week, that came to 100 marks per year. In 1926, Einstein described the conditions as *schwierig*, his way of saying, "Only a dog should live that way." With Einstein's help, Kaluza finally obtained a professorship at the University of Kiel in 1929. He moved to Göttingen in 1935, where he became a full professor. He remained there until his death nineteen years later. Not until the 1970s was the possibility of new dimensions looked at seriously again.

34. The Birth of Strings

HO KNOWS WHEN inspiration will come? Even harder to know is where it will lead. The story of string theory begins atop a mountain 750 feet above the Mediterranean. The town is Erice, Sicily—a slow, hot town of narrow streets and ancient stone. Erice was Erice when Thales roamed the earth. Today the town is defined largely by its Centro Ettore Majorana, a cultural and scientific center in which a series of "summer schools," roughly a week long, have been held each year for decades. The Ettore Majorana schools are gatherings where graduate students and young faculty gather with leaders in the field, and attend lectures on cutting-edge topics in their field.

In the summer of 1967, one such cutting-edge topic was an approach to elementary particle theory known as S-matrix theory. Gabriele Veneziano, an Italian graduate student from the Weizman Institute in Israel, was sitting in the audience, listening to an intellectual hero, Murray Gell-Mann. Gell-Mann would soon receive the Nobel Prize for his discovery of the concept of quarks, then thought to be the inner constituents of the family of elementary particles called *hadrons* (which include the proton and neutron). The inspiration Veneziano would have there would, within a few years, launch string theory. Gell-Mann's topic: the regularities of a mathematical construct called the *S-matrix*.

Invented by Heisenberg, the S-matrix approach was introduced by John Wheeler in 1937 and championed in the 1960s by a Berkeley physicist named Geoffrey Chew. The "S" stands for "scattering" because that is the main way physicists study elementary particles: they speed them up to enormous energies, smash them into each other, and look at the remnants that fly out. It is like studying the automobile by setting up car crashes.

In tiny crashes you might detach something boring like a bumper, but at race car speeds the nuts and bolts screwed deep within the passenger seat could fly within an experimenter's keen sight. There is one big difference, though. In experimental physics, smashing a Chevy into a Ford could result in an explosion of parts from a Jaguar. Unlike autos, elementary particles can morph into each other.

When Wheeler developed the S-matrix, there was a growing body of experimental data, but no successful quantum theory of particle creation and annihilation, not even for electrodynamics. The S-matrix was a black box that took input—the colliding particles' identities, momenta, etc.—and created as output the same kind of data, but for the emerging particles.

To *construct* the S-matrix—the insides of the black box—in principle you need a theory of the interaction. But even without a theory there are certain things you can say about the S-matrix based only on the symmetries of nature and general principles, such as requiring consistency with relativity. The crux of the S-matrix approach was to see how far you could go on these principles alone.

In the fifties and sixties, this approach was a bit of a rage. In his Erice talk, Gell-Mann spoke of some striking regularities called *dualities* that were observed in the collisions of hadrons. Veneziano wondered if these regularities might occur in more general circumstances. It took him a year and a half, but Veneziano finally realized this: all the mathematical properties of the S-matrix he sought were possessed by a single, simple mathematical function called the *Euler beta-function*.

Veneziano's theory, dubbed the dual resonance model, was

a striking discovery. Why should the potentially complex S-matrix take on such a simple and graceful form? It was the first of the mathematical miracles that would crop up regularly in string theory, the type of beautiful result that would convince Schwarz he wasn't wasting his life pursuing it.

Veneziano's result was so elegant it inspired physicists to ask a decidedly non-S-matrix question: What are the details of the collision process that produce this S-matrix? What is inside the black box? If they figured it out, they would be elucidating the internal structure of the colliding hadrons, and the force, called the *strong* force, that governs them.

In 1970, Yoichiro Nambu of the University of Chicago, Holger Nielsen of the Niels Bohr Institute, and Leonard Susskind, then at Yeshiva University, answered the question: You model the fundamental particles not as points, but as tiny vibrating strings.

Do you *discover* a theory or *invent* one? Are physicists kids with flashlights searching the park at twilight for traces of truth, or are they kids with blocks trying to build structures high before they topple? Or in truth can the process be both—a duality like the one Gell-Mann was talking about or the one between particles and waves?

There are less kind words for both *invent* and *discover.* Like *concoct* and *stumble upon.* The original string theory—called *bosonic string theory*—was certainly a concoction. It was artificial, had many unreal features, and was clearly put together simply to reproduce Veneziano's insight. But Nambu et al. had also stumbled upon something. They had discovered string theory in much the same sense that Planck discovered quantum theory. They both had found an idea—that energy levels can be quantized, or that particles might be modeled as strings—whose meaning and scope were not understood, and which would take years to develop into a meaningful theory. They had stumbled across something that might be a new principle of nature, or simply a mathematical trick. Only years of effort could determine which. In the case of quantum theory, it was twenty-five years from Planck to Heisenberg and Schrödinger. String theory has already passed that benchmark.

35. Particles, Schmarticles!

DECADE before strings, Geoffrey Chew, one of the most promising physicists of the 1960s and late 1950s, stood up at a conference and declared that field theory was no good. There should be no elementary particles, said Chew. We should think of particles as being composed of each other. He proposed that physicists look for a kind of one-particle-makes-them-all theory dubbed, in the spirit of the cold war, nuclear democracy. Moreover, Chew didn't believe in the approach of developing different theories based on and fitted to the properties of the various different forces. He believed that if physicists examined all possible S-matrices closely enough, they would find that only one was consistent with general physical and mathematical principles. That is, he believed that the universe is as it is because this is the only way it could possibly be.

Today, we know that the conditions Chew imposed were not enough to completely specify the physics. Witten calls S-matrix theory "an approach, not a theory." Gell-Mann says it was overblown, a pompous name for an approach he himself first presented at a conference in Rochester, New York, in 1956. Yet, says Gell-Mann, "the S-matrix approach was the right approach. It is still used today for string theory." There was good reason for Chew to hold these aesthetics. Even today's standard model, despite all its successes, is not pretty. The problems began in 1932, when two new and exotic particles were discovered. One was the positron, the anti-particle of the electron. The other was a new member of the nucleus that is quite like a proton but carries no electric charge—the neutron. Physicists were reluctant to accept the possibility of new particles. Other explanations were concocted. Dirac, whose theory predicted the positron, was cowed into calling it at first a kind of proton-lite (the positron has the same charge as a proton but less than $\frac{1}{1000}$ the mass). Attempts were made to explain the neutron as a proton and electron hugging really tight. But, like parents of a teen-aged child, it was difficult for physicists to hold their ground. Soon physicists admitted not only the new particles but also the concept of anti-matter and two new forces, the strong force and the weak force, important within the atomic nucleus.

By the 1950s, particle accelerators allowed the study of dozens of new particles—neutrinos, muons, pions . . . J. Robert Oppenheimer suggested that the Nobel Prize go to the physicist who *didn't* discover a new "elementary" particle. Enrico Fermi remarked, "If I could remember the names of all these particles, I would have become a botanist."

Physicists coped with all this change by developing new theories called quantum field theories to describe how particles appear and disappear. Quantum mechanics had been designed to describe situations in which particles interact, not those in which they are created, destroyed, or transformed into one another. In a quantum field theory, there is only one way that anything in the universe ever interacts: by exchanging particles known as messenger particles. What physics had for centuries called "force" is, according to field theory, just a higher-level description of the exchange of particles between other particles.

Think of two basketball players running down the court, passing the ball back and forth. They are the particles in question. Their interaction, whether it brings them closer or pushes them away, is carried by the ball, the messenger particle. For electromagnetism, the messenger particle is the photon. In quantum electrodynamics, charged particles like the electron and proton feel the electromagnetic force through the exchange of photons. Uncharged particles like the neutrino do not exchange photons.

The first successful quantum field theory was that of the electromagnetic field, developed in the 1940s by Feynman, Julian Schwinger, and Sin-itiro Tomanaga. In the 1970s, a new theory was created uniting the theory of the electromagnetic field with one of the weak force. Soon, in analogy to quantum electrodynamics, a theory for the strong force was invented, with its messenger particles, the *gluons*. Collectively, the field theory of these three forces is what comprises the standard model.

Physicists had done an admirable job—if you're a botanist. The classification of the elementary particles in the standard model, though a triumph in predictive power, is not pretty. For example, the elementary particles of matter—as opposed to the messenger particles—come in families. Each family contains four particles—an electronlike particle, a neutrinolike particle, and two quarks. One of these families contains the ordinary electron and neutrino, and the two quarks that constitute the familiar proton and neutron. The corresponding particles in the other two families differ only in their mass—with each "exotic" family containing successively heavier particles. The standard model mirrors this structure, but it is incorporated into the model without explanation. Why are there three families, and why four members of each? Why are the masses what they are? The standard model gives no insight into these questions.

The strength of each force is also input without explanation, encoded in numbers called *coupling constants*. The response of a particle to a force is characterized by a quantity called *charge*—a generalization of the electric charge. Typically, a given particle carries more than one type of charge—that is, it feels more than one type of force. These charges are also unexplained inputs of the theory.

If Fermi had a problem remembering the names of the elementary particles, the standard model would only have made things worse. To remember its equations, he would have to remember the values of nineteen underived parameters. Not nice numbers that would have made Pythagoras proud, but nasty numbers with names like the Cabibbo angle, and values like 1.166391×10^{-5} (the Fermi coupling constant in GeV^{-2}). The book of Genesis says, "Let there be light: and there was light." According to modern physics, God also carefully adjusted the fine structure constant to precisely 1/137.035997650 (give or take a few parts per billion).

Without venturing into the philosophy of science, there is something about the phrase "fundamental theory" which seems to imply that dozens of researchers should not be making their livings measuring its nineteen "fundamental" parameters to accuracies of seven decimal places. You feel like tapping the theorists on the shoulder and asking, "You ever hear of a guy named Ptolemy?" With enough circles upon circles, a clever scientist could match any data.

String theorists rebel against the idea that this model is fundamental. From their theories, they hope they will some day be able to derive it. Like S-matrix theorists, but unlike field theorists, they aim to not have to specify input parameters—not even structural ones such as the number of dimensions of space. Like Chew, their aim is to find a theory completely defined by general principles. They hope that from it, they can understand the origins and strengths of all forces, the types and properties of particles, the structure of space itself. And in their theory, as in Chew's dream, one particle fits all. The difference is, in their theory, the particle is a string.

This string is made of nothing, for to define a material composition implies a finer structure that they do not possess. Yet everything is made of them. At a length of 10^{-33} centimeters, they are protected from our direct observation by a factor of 10^{16}. On the eye chart, they may be oriented vertically, horizontally, along the diagonal. But even in our most microscopic peering, today's technology flunks the test. "Down? Up? Sideways? . . . Sorry, Doc, all I really see is a bunch of dots."

That strings are hidden by their minute size should not be a surprise—after all, they were theorized, not observed. But the level of their hiddenness can only be termed overkill. It has been estimated, variously, that the accelerator needed to directly detect one experimentally would be somewhere between the size of our galaxy and that of the entire universe. A historian digging up a withered copy of this book in the year 3000 might chuckle at the estimate, for by then we may have learned to see them by mixing vermouth and vodka (in just the right proportion). In the meantime, direct observation seems out of the question.

In quantum mechanics, waves and particles are dual aspects of the same phenomenon. In quantum field theory, both matter and energy particles are considered to be excitations of various quantum fields. This is true also in string theory, but in string theory there is just one field. All particles arise as vibrational excitations of just one type of elementary object: the string.

Picture a guitar string that has been tuned by stretching it with the proper tension. The musical notes of the string are called *excitation modes* compared to the string at rest. In acoustics, they are known as higher harmonics. In string theory, they present themselves as different particles.

It was the Pythagoreans who first studied the mathematical and aesthetic properties of musical sounds. They discovered that when you tweak a string it vibrates with a pitch, or frequency, that varies inversely with the length of the string. This fundamental frequency comes from the mode of vibration in which the maximum displacement of the string occurs at its midpoint. But the string can also vibrate in a way such that its midpoint never moves, and maximum displacements occur halfway between the midpoint and each end. This would be the fundamental mode of vibration if you held the string down at its midpoint. It is a vibration with two equivalent waves within the string length, that is, with half the wavelength and double the frequency of the fundamental. It is called the second harmonic, in musical terms, and it is an octave higher.

A pluck of the string will also produce vibration with the form of three complete waves, four waves, and so on (but never a fractional number of waves, for this would violate the condition that the end of the strings are fixed). These are the higher harmonics. A note on the violin and piano, for instance, generally is accompanied by a stronger relative amplitude of the first six harmonics than are present for other instruments. The sound of an organ pipe, on the other hand, is generally comparatively poor in the upper harmonics. The higher harmonics are what give musical instruments, and the families of elementary particles, their variety.

The strings of string theory are not tied down like guitar strings. They can be open or closed. They can split and rejoin or fuse at their ends to form a loop or fuse and split to form two loops. As a string splits or joins, its properties change—from a distance it looks as if it is a new type of particle. The exchange of messenger particles is really the splitting and joining of strings floating in space-time.

It is as if the different particles we observe are music boxes, their properties the notes we hear them play. Categorized by their music, there seem to be many different classes of music box. According to string theory, the music boxes are all physically identical, differing not in their makeup, but in the way the string within them vibrates.

For instance, the energy of vibration depends on its wavelength and amplitude. The more peaks and troughs within its length, and the greater their size, the more energetic the vibration. Since we know from relativity that mass and energy are equivalent, it is probably not surprising that, seen from outside the black box, strings vibrating more energetically are perceived by us as more massive.

This is also true of properties other than mass, such as the various types of charge. And why not? In the sense of field theory, a particle's mass is just a kind of charge—its charge with respect to the gravitational force. According to string theory, all the particles of nature, including the messenger particles, with all their diverse spectrum of properties, are simply different patterns of vibration of the string.

There is a great variety and complexity of particles in the universe. Is there richness enough in a vibrating string to encompass all their diversity? Not in Euclid's world.

But the modes of vibration of a string, and therefore the predictions of which particles exist, and their properties, depend greatly on the number of dimensions in which the string vibrates, and on the topology of the dimensions. This is the source of the deep connection between the properties of space and the properties of matter itself; according to string theory, the structure of space determines the physical properties of the elementary particles and forces of nature. In string theory, just three space dimensions just won't do. It is the precise geometry and topology of the extra dimensions that determine the theory of elementary particles and forces that string theory predicts.

A string in one dimension can only vibrate in one way—it can stretch and compress. This kind of vibration is called a longitudinal vibration. In two dimensions, a string can still vibrate in this way, but it has a completely new type of vibration open to it—the transverse vibration, in which its motion is in a direction perpendicular to its length. These are essentially the vibrations we have just discussed. In three dimensions, the direction of the transverse vibration can rotate or spiral—just picture a Slinky. In higher dimensions the complexity increases.

Topology also affects the vibration. Topology is a hard subject to define, but roughly, it deals with the properties of surfaces and spaces that relate to their shape but not to their metric (distance relations) or curvature. A line segment is topologically different from a circle because it has two ends, while a circle has none. Yet the difference between a circle and an ellipse does not interest the topologist—it is only a matter of curvature. One way of thinking of these distinctions is this: any two shapes that can be transformed into each other by stretching but not tearing have the same properties as far as the topologist is concerned.

How does the topology of space affect the string? Suppose that string theory had called for only two extra dimensions. Since the extra dimensions in string theory are supposed to be small, imagine a "small" two-dimensional space—a square or a rectangle—like the plane, but finite. That space has one type of topology. Now imagine rolling it up into a cylinder. Though intuitively it may seem curved, geometrically, a cylinder is considered flat like a planar space. That means it has zero curvature: any figure you draw on the plane can be rolled up into the cylinder without distortion of the distance measurements between any two points. But the cylinder does differ from the plane in its connectedness, or topology. For instance, on a plane, any circle or other simple closed curve can be shrunk down to a point without leaving the plane. On a cylinder, there are curves for which this cannot be done—for instance, any curve that circles the axis of the cylinder. The vibrational motion of this type of string on a cylindrical space, being constrained, is different from that on a plane, so in string theory, such a universe would result in different types of particles and forces. The cylinder is closely related to another shape, the torus, or donut. To get the torus from the cylinder, just connect the ends. But far more complex topologies are possible: for instance, instead of a donut with one hole, you can have a donut with multiple holes. Each of these yields still different vibrational spectra. The more dimensions we add, the more complex the possible spaces, especially if we allow spaces that are not flat. And in all these different spaces, the possible modes of vibration are different. This richness of types of vibration is what allows string theory to account for the variety of elementary particles and forces—at least in theory.

At this point it would have been nice to be able to say that due to various consistency requirements, there is only one type of space possible for the extra dimensions of string theory, and that the properties of elementary particles that correspond to the string vibrations in that space are exactly those we observe in nature. Dream on. There is some good news, though. For one, not just any extra dimensions will work. It appears there must be six (a point we will come back to later), and that they must have certain properties, such as being curled up like the extra dimension in Kaluza's theory. In 1985, physicists discovered the class of spaces that has just the right properties. They are called Calabi-Yau spaces (or Calabi-Yau shapes; they are, after all, finite spaces). As one might guess, six-dimensional Calabi-Yau spaces are more complicated than, say, a chocolate donut. But they do have one thing in common—a hole. Actually they may have various numbers of holes, and even the holes are complicated, multidimensional objects, but those are technical details. The point is this: there is a family of string vibrations associated with each hole. String theory thus predicts that elementary particles come in families. It is an example of one of the striking "derivations" of experimentally observed facts that in the standard model had to be incorporated "by hand," without theoretical explanation. That's the good news.

The bad news is that there are tens of thousands of known types of Calabi-Yau spaces. Most contain more than three holes, though there are only three families of elementary particles. And to attempt the calculations needed to derive the properties of particles that the standard model merely proclaimed, for instance, their mass and charge, physicists need to know which of the many possible spaces to employ. So far, no one has been able to find the Calabi-Yau space that yields the precise description of the physical world as we know it, that is, the standard model, or a fundamental physical principle that would justify choosing one space over another. Some are skeptical that the approach will ever bear fruit. But the critics are far fewer and vastly quieter than they were in the beginning, when, for many years, to work on string theory was the kiss of professional death.

36. The Trouble with Strings

HEN NAMBU et al. proposed string theory, it had some peculiarities. For one, their theory was not consistent with relativity unless a certain unfriendly factor could be made to equal zero: $[1 - (D-2)/24]$. Any high school kid can tell you the answer to this one: $D = 26$. But that is only the beginning of the problem, for D in this equation represents the number of dimensions of space. Soon interest would be revived in Kaluza's work, only now his five dimensions would seem not too many or too weird, but not weird enough.

The theory had other problems. As mentioned above, when the probabilities for the occurrence of certain processes were calculated according to the rules of quantum mechanics, the mathematics gave negative numbers. The theory also predicted the existence of certain particles called *tachyons* whose mass was not a real number, and which traveled faster than light. (Einstein's theory doesn't strictly speaking prohibit this; it only prohibits particles from moving *at* the speed of light.) And it predicted certain other extra particles, never observed.

If your local weather forecast predicted a negative 50 percent chance for thundershowers, with rain falling up, and frogs falling from the sky, you probably wouldn't have much faith in their computer model. Physicists, too, were skeptical. But suppose the forecast also foretold the temperature, and it got that right. The match between bosonic strings and hadron behavior was too intriguing to ignore.

If all this seemed awkward, physicists soon realized that the theory had another fault, and for the theory, this one was *really* embarrassing. In quantum mechanics, all particles belong to one of two types: *bosons* and *fermions.* On the technical level, the difference between bosons and fermions is a type of internal symmetry known as spin. But on the practical level, the difference is this: no two fermions can occupy the same quantum state. That is a good property if you are building up, say, the atoms of matter. It means that the electrons in the atom don't all congregate in the lowest energy state. If they did, the electrons of all the elements would have a tendency to remain in their lowest energy state. Instead, the atoms of the periodic table are built by filling, one by one, the outer electron states, endowing the elements with their physical and chemical differences. The bosons have no such restriction. Thus, matter is made of fermions. The messenger particles, involved in transmitting forces, they are the bosons. But in bosonic string theory, *all* the particles are—guess what?—bosons.

This is the problem with string theory that Schwarz first attacked. It won him his mentor, and his chance to remain at a top university where his work, if not believed, would at least be heard.

In 1971, Pierre Ramond of the University of Florida derived a string theory for fermions by discovering an early form of a new symmetry called supersymmetry, which connects bosons and fermions. Then, with André Neveu, Schwarz developed a theory known as spinning string theory that included both fermionic and bosonic type particles, eliminated the tachyons, and reduced the number of dimensions required from twenty-six to ten. Their work proved to be a major turning point in string theory, and in Schwarz's career.

Gell-Mann, who was working at CERN (the European Laboratory for Particle Physics) in Geneva at the time, says, "As soon as Schwarz's paper came out, I hired him." They hadn't even met. The next fall, Schwarz moved to Caltech from Princeton, where Schwarz had just been denied tenure.

While Feynman lumped string theory with the other miracle cures that had come and gone over the years, Gell-Mann shared Schwarz's belief in it. "It had to be good for something," he says. "I didn't know what, but something." In 1974, Gell-Mann also brought another string theorist to Caltech for a visit, Joel Scherk. Schwarz and Scherk soon made an astounding discovery.

String theory contained a particle that had the properties of the gluon, the messenger for the strong force. But one of its embarrassments was an extra particle—a messenger-type particle that didn't seem to have any relevance. Until the work of Schwarz and Scherk, the length of the string had been assumed to be around 10^{-13} centimeters, roughly the diameter of the hadron. But they found that if instead you assumed it was much smaller, 10^{-33} centimeters, the Planck length, the extra messenger particle fit exactly the properties of the graviton, the hypothesized messenger particle of the gravitational force. String theory wasn't just a theory of hadrons—it included gravity, and perhaps the electroweak force as well!

But wait; haven't we learned that mixing gravity and quantum mechanics leads to chaos and contradiction? In Schwarz and Sherk's theory, because the strings were not dimensionless points but objects with a finite length, the problems in the ultramicroscopic realm did not arise. They had found what they thought was a consistent quantum field theory from which they could derive Einstein's equations, but which, on the ultramicroscopic scale, behaved differently in just the way required to avoid contradictions between general relativity and quantum mechanics. Einstein, on publishing relativity, had expected to be attacked. Schwarz and Scherk expected a torrent of excitement.

Schwarz and Scherk traveled the world giving lectures. People applauded politely, then ignored their work. If pressed, they said they didn't believe it. In defense of these "people," the mathematics was (and still is) extraordinarily hard and complex. "People didn't want to make the investment to understand it, and without the imprimatur of a statesman, they wouldn't make the effort," says Schwarz.

Gell-Mann would have qualified as just such a statesman, but he himself did little research in the field. The few papers he did with Schwarz, chuckles Schwarz, "were among both of our most forgettable." There was no professorship at Caltech for John Schwarz. Just a series of extensions of his research position. "I couldn't get John a regular academic job," says Gell-Mann. "People were skeptical." In 1976, Scherk and others showed how to incorporate supersymmetry into string theory, finally creating the theory called superstrings. It seemed to be another groundbreaking result, but no one seemed to care. They were more interested in a competing theory called supergravity, and in more traditional quantum field theory sans gravity, the standard model. Uniting the electromagnetic force with the weak and strong nuclear forces, the standard model was achieving one triumph after another, including the experimental creation in 1983 of the W and Z bosons, the messenger particles of the weak force.

String theory hit a long dry spell. No one knew how to do any practical calculations using the theory. The extra dimensions and other problems remained. Meanwhile, Joel Scherk suffered a breakdown. He would be found crawling around the streets of Paris. He sent strange off-the-wall telegrams to physicists such as Feynman. He still managed to work, at least part of the time, amazing his doctors—and his colleagues. Then he split up with his wife, who moved to England with their children. In 1979, he committed suicide, a great loss to the small band of string theorists. In the early 1980s, new problems with string theory were found. Schwarz seemed to most to be stuck on a blind alley, nothing ahead of him but a dead end.

It was noted by some that he was imitating the "wasted" effort of the man who had been his doctoral adviser, Geoffrey Chew. With a goal similar to Schwarz's, Chew had spent twenty-five years working on S-matrix theory. The first few he spent in good company; the latter fifteen years he worked virtually alone, like Schwarz, the subject of occasional derision. In the end, Chew gave up on his dream. In retrospect, Chew's efforts were not for naught: says Schwarz, "it is not clear that without him there would have been string theory. It grew out of the S-matrix approach."

At Caltech, through it all, Gell-Mann remained a powerful booster. "It made me happy and proud we had them [Schwarz and Scherk] at Caltech," he says. "It was really heartwarming. I had a gut feeling. So at Caltech I maintained a nature reserve for endangered species. I had done a lot of conservation work in the third world. Here I was doing it at Caltech." In 1984, Schwarz made another breakthrough, this time working with Michael Green (then at Queen Mary College in London). They found that in string theory, certain unwanted terms that might lead to anomalies miraculously cancel each other. The result was presented at a workshop in Aspen that summer, dramatized in a skit at the Hotel Jerome. It ended with Schwarz being carried off the stage by men in white coats while he yelled that he'd found a theory of everything. The sardonic humor of the skit reflected his expectations— that this result, too, would be dismissed and ignored.

But this time, before Schwarz and Green could finish writing up their result, a fellow named Edward Witten called. He had heard about their talk from others at the workshop. Schwarz was delighted to have any new interest in his work. But Witten wasn't just a researcher won over. He was the most influential physicist and mathematician in the world.

Within a few months, Witten (then at Princeton, currently at Caltech with Schwarz) and his co-workers had achieved several new major results, such as the identification of the Calabi-Yau spaces as candidates for the curled-up dimensions. That's all it took to convince hundreds of physicists to start working on string theory. Schwarz had finally achieved the imprimatur he needed.

Schwarz suddenly had interest from other top universities, anxious to lasso the newly great scientist. Gell-Mann was determined to finally win him tenure. Even then, it wasn't easy. One administrator commented, "We don't know if this man has invented sliced bread, but even if he has, people will say that he did it at Caltech, so we don't have to keep him here." But after twelve and a half years, Schwarz did get tenure. It was several years quicker than Kaluza.

Today, Schwarz's paper with Green is defined as "the first superstring revolution." Witten says, "Without John Schwarz, string theory would very probably have become extinct, perhaps only to be rediscovered some time in the twenty-first century." But the baton had been passed. A decade later, Witten would dominate, and eventually engineer his own revolution in string theory.

37. The Theory Formerly Known as Strings

Y THE early 1990s, string theory had cooled down in popularity. A few years earlier, the *Los Angeles Times* had gone as far as endorsing the position of one critic who wondered whether string theorists should be "paid by universities and be permitted to pervert impressionable young students." (Hopefully, these days, the *LA Times* sticks to issues closer to its local expertise, like how it's going between Warren Beatty and Annette Bening.) There were good reasons for the excitement to wane. Lamented string theorist Andrew Strominger, "There are some big problems." Part of it was the lack of more startling new predictions squeezed from the theory. But there was also a new embarrassment—one every bit as bad as the ones in the old days. There seemed to be five different kinds of string theory. Not five different Calabi-Yau candidates—only five of *those* would have been good news—but five fundamentally different structures for the theory. To paraphrase Strominger, it is unaesthetic to have five different unique theories of nature. The dry spell would last ten years. It was another long desert for Schwarz to cross. But this time he had plenty of company in his search for the promised land, and a prophet to lead him.

Every generation in physics has its dominant figures. In the decades before string theory, it was Gell-Mann and Feynman. For the last few decades it has been Edward Witten. Says Brian Greene of Columbia University, "Everything I've ever worked on, if I trace its intellectual roots, I find they end at Witten's feet." I first heard of Witten in the late seventies, as a physics major at Brandeis University, where he had preceded me by a few years. I was treated to a couple of professors' re-

marks to the effect of "You're smart, but you're no Ed Witten."
I wondered, would the same professors have said to their
wives, "You're good, but my old girlfriend was really, *really*
good"? When I thought it over, I decided I could imagine them
saying that. But still, I wanted to know, who was this genius?

To my chagrin, it turned out he had majored in *history,* one
of those non-science fields we physics majors thought had the
intellectual depth of high school except with more reading
homework. Worse yet, he hadn't taken a single physics
course. Apparently, the physics in which he so hopelessly
outshone me was, for this Einstein, just a hobby.

I was happy to discover that Witten had worked on the Mc-
Govern campaign in '72, meaning that while he may have
been laudably anti-Nixon, he was hopelessly challenged in
the area of "a good use of one's time." And, if he was such a
genius, how come George McGovern didn't win? But Mc-
Govern *had* won in Massachussetts—the only state in which
he had. Could that have been due to the work of Witten? A
few years ago, I learned it wasn't. McGovern, tracked down
in his retirement by a reporter anxious to know his opinion of
the "smartest man in the world," answered that he didn't remember Witten. Then he agreed with the assessment anyway,
saying, "Well, he was smart enough to back McGovern in
'72, and I judge everybody by that criterion."

After Brandeis, Witten ended up a graduate student in
physics at Princeton. Not having ever taken a physics course,
he had no qualifications for entry, but it turns out they had a
special admissions program for kids destined to become the
smartest person in the world. When I finally met Witten, I was
a graduate student myself, at Berkeley, where, before accepting me, they had no doubt gone over my grades and other
qualifications, earned in *actual* physics courses, with a fine-
toothed comb.

Witten turned out to be a tall, gangly guy, with black hair, and glasses with black plastic frames. He was intense, but nice enough, and so soft-spoken you had to squint with your ears to make out what he was saying. (It usually proved to be worth it.) He stopped in the midst of his talk that day, apparently to think some deep thoughts. But he was silent so long that people started to clap, like the ignoramuses at a Beethoven concert who mistake the end of a movement for the finale. Witten told us, sounding somewhat annoyed, that his symphony wasn't over.

Today, Witten is often compared to Einstein. There could be many reasons for that, but the foremost one is probably that those making the comparison haven't heard of many other physicists. It's really the curse of Einstein's legendary status—he's become a cliché, everyone known as the Einstein of this or the Einstein of that. That's what you get for being the Cadillac of physicists. There are superficial similarities between Einstein and Witten. They are both Jewish, spent years at the Institute for Advanced Study, and showed a strong interest in Israel and an attraction to peace movements. At the age of fourteen, Witten's letters to the editor opposing the war in Vietnam were getting published by his local paper, the *Baltimore Sun*, and he has been involved with peace groups in Israel.

But if you must make a comparison, in his work Witten is far more like Gauss than Einstein. Not dependent on any old friend to explain modern geometry to him, like Gauss, Witten has been reinventing it himself. And like Gauss, his work is having a major impact on the direction of modern mathematics, something that Einstein's work never did. And then there's the flip side—Witten's (and everyone else's) approach to string theory, and now M-theory, is driven by insights of mathematics, not physical principles as Einstein's was. Not by choice, perhaps, but by historical accident: the theory was after all stumbled upon. The new principle of physics that is at its core, a kind of "happiest thought" for Witten, is, if it exists at all, yet to be discovered.

In March 1995, Edward Witten spoke at a string theory conference at the University of Southern California. It had been eleven years since Schwarz's superstring revolution, and to many people string theory seemed to be slowly unraveling. Witten's talk changed everything. What he explained was another mathematical miracle: all five of the different string theories, he asserted, are merely different approximate forms of *the same* grander theory, now called M-theory. Physicists in the audience were floored. Nathan Seiberg of Rutgers University, due to speak next, was so awed by Witten's talk that he remarked, "I should become a truck driver."

Witten's big breakthrough is now known as the second superstring revolution. According to M-theory, strings are not really the fundamental particle, but are only examples of more general objects, called *branes,* short for membranes. Branes are higher-dimensional versions of the string, which is a one-dimensional object. A soap bubble, for instance, would be a two-brane. According to M-theory, the laws of physics depend upon the more complex vibrations of these more complex entities. And, in M-theory, there is an extra curled-up dimension—making a total of eleven dimensions, not ten. But its strangest aspect is this: in M-theory, space and time, in some fundamental sense, do not exist.

M-theory appears to have the property that what we perceive as position and time, that is, the coordinates of a string or a brane, are really mathematical arrays known as matrices. Only in an approximate sense, when strings are far apart (but still close on the scale of everyday life), do the matrices resemble coordinates—because all the diagonal elements of the array become identical and the off-diagonal elements tend toward zero. It's the most profound change in the concept of space since Euclid.

Witten used to say that the M in M-theory stood for "mystery, or magic or matrix, my three favorite words." Lately, he has added the word *murky,* presumably not a favorite word. M-theory is yet more difficult to understand than string theory. No one knows what equations would arise from it, much less how to approximate their solutions. In fact, not much is known about it at all, except that it seems to exist—a grander theory to which the five types of string theories are merely various different types of approximation. Yet the ideas of M-theory have led to the most striking indication yet that there is something to this string idea: a prediction having to do with the physics of black holes.

Black holes are one of the phenomena predicted by general relativity. Their defining feature is that they are black (which to a physicist means that no light or radiation can escape them). In 1974, Stephen Hawking said, Aaaggghhhh—wrong answer! If one considers the laws of quantum mechanics, one is forced to conclude that black holes are not really black. That is because, due to the uncertainty principle, empty space is not really empty; it is filled with pairs of particles and antiparticles that exist for the tiniest moment before annihilating each other back into oblivion. According to Hawking's very involved calculations, when this happens in the space just outside the black hole, the black hole can suck in one member of the pair, shooting the other member out into space, to be observed as radiation. Hence, black holes glow. That also means that they have a non-zero temperature, just as the glow from coals indicates the presence of a certain amount of heat. Unfortunately, the temperature of a typical black hole would be less than a millionth of a degree, too low to be observed by astronomers. But for physicists, the realization that black holes had a temperature at all led to a really amazing conclusion. If black holes have temperature, they can have something called entropy, and in fact, the amount of entropy they would have is enormous—written as a number, it would occupy more than a line of text in this book.

Entropy is a measure of the disorder in a system. If you know a system's internal structure, you can calculate its entropy by counting the number of possible states it can be in: the more states, the higher the entropy. For instance, if Alexei's bedroom is cluttered, it has many states accessible to it—the hamsters could be here, the pile of dirty clothes there, the old comic books somewhere else, or all the items could be rearranged, forming a different "state." The more junk in his room, the more possible states (contrary to popular belief, a condition of high entropy has nothing to do with any particular neat or messy arrangement, just the total number accessible to the system). But if his room were empty, it would have only one state accessible to it—there would be nothing to rearrange—and its entropy would be zero. Before Hawking, black holes, thought to have no internal structure, were thought to be something like an empty room. But now it seems they are like Alexei's actual room. Had Hawking asked, I could have confirmed this: I have always told Alexei that his room was like a black hole.

Physicists puzzled for two decades over Hawking's result. Combining the separate theories of relativity and quantum theory is a tricky business. Where are all the states of the black hole that this entropy refers to? No one knew. Then, in 1996, Andrew Strominger and Cumrun Vafa published a spectacular calculation: employing the ideas of M-theory, they showed that you could create certain types of (theoretical) black holes out of branes; for these black holes, the states are brane states—and you can count them. The entropy they calculated in this way agreed with the entropy that Hawking predicted employing his completely different method.

It was striking evidence that M-theory is doing something

right, but still, really another postdiction. What the theory needs, those irritating experimentalists insist on reminding us, is some confirmation from the world of reality. Hope for experimental evidence for M-theory currently lies in two areas. One is the possible discovery in the next decade of supersymmetric particles. This could happen at the new Large Hadron Collider (LHC) at CERN in Geneva. The other test will be a search for deviations in the law of gravity. According to Newton, and on this scale, also Einstein, two lab-sized objects should attract each other with a force proportional to the inverse of the square of their separation—halve the distance between them and their attraction becomes four times as strong. But, depending on the nature of the extra dimensions, it is possible in M-theory that as objects get extremely close, their attraction will increase much faster. And though physicists have probed the behavior of other forces down to a scale of almost 10^{-17} centimeters, they have so far studied the behavior of gravity only at distances greater than about 1 centimenter. Researchers at Stanford University and the University of Colorado at Boulder are currently conducting experiments employing "desk-top" technology to test gravity at smaller distances.

Schwarz isn't worried. He says, "I believe we have found the unique mathematical structure that consistently combines quantum mechanics and general relativity. So it must almost certainly be correct. For this reason, even though I do expect supersymmetry to be found, I would not abandon this theory if supersymmetry turns out to be absent."

Nature evolves with hidden order. Mathematics reveals it. Will M-theory be the beautiful textbook theory of tomorrow's college physics course, or a footnote in the history of science lecture entitled "Dead Ends"? Whether Schwarz is Oresme and Witten Descartes, or whether together they play the role of Lorentz, building a hopeless mechanical theory out of the nonexistent ether, has not yet been revealed. As a young scientist Schwarz knew only that this theory was too beautiful not to be good for something. Today, a whole generation of researchers looks at nature and sees his strings. It would be hard to view the world the old way ever again.

EPILOGUE

As kids we play with puzzles; as human beings we live in one. How do the pieces fit together? It is a puzzle not for the individual but for the organism called humanity. Are there really laws of nature? How do we come to know them? Is natural law a hodgepodge of local statutes, or is there a unity in the universe? To the human brain, that humble gray blob that still too often stumbles over such "simple" subjects as love and peace, or the cooking of a good risotto, the hugeness and complexity of the cosmos ought to be abstruse beyond imagination, impossible to conceive. Yet, for over 100 generations we have been piecing it together.

As human beings, we naturally seek order and reason in the workings of the world around us. Our tools we inherited from the ancient Greek geometers, who not only gave us the exact reasoning of mathematics, but also taught us to look for the aesthetic in nature. They found satisfaction in the roundness of the sun and earth and in the orbits of the planets, for to them the circle and the sphere were the most perfect of shapes. After the Dark Ages, with the revival of Euclid's *Elements* and the birth of the experimental method, we found that order extends beyond the what of nature to the why of natural law. Experiments in the seventeenth century showed that all bodies fall at the same rate regardless of their makeup, size, or weight, or whether it is Galileo dropping them or fellow experimenter Robert Hooke. Observation has since confirmed that the same laws that govern the attraction of the earth to Newton's apple also apply to the moon, or to the motion of distant planets around their own stars. And these laws appear to have survived, unchanged, since the beginning of

263

time. What power imposes on the universe that all things follow certain special rules? And why don't the laws change over time, or from place to place, over billions of years and trillions of miles? It is not hard to understand why some people have always found the answer in God. But the course of science is the one set by the Greek geometers, and mathematics was their tool. Ever since the Greeks, mathematics has been at the heart of science, and geometry at the heart of mathematics.

Through Euclid's window we have discovered many gifts, but he could not have imagined where they would take us. To know the stars, to imagine the atom, and to begin to understand how these pieces of the puzzle fit into the cosmic plan is for our species a special pleasure, perhaps the highest. Today, our knowledge of the universe embraces distances so vast we will never travel them and distances so tiny we will never see them. We contemplate times no clock can measure, dimensions no instrument can detect, and forces no person can feel. We have found that in variety and even in apparent chaos, there is simplicity and order. The aesthetics of nature reach beyond the grace of the gazelle and the elegance of the rose, out to the farthest galaxy and into the tiniest crevice of existence. If current theories prove valid, we are nearing the great epiphany of space, an understanding of the interplay of matter and energy, space and time, the infinitesimal and the infinite.

Is our understanding of physical law truth, or is it merely one of many possible descriptive systems? Is it a reflection of the universe, or of our own innate viewpoint as a species? It is one miracle that the regularities in physical law exist, another that we can discern them, but the greatest miracle of all would be if our theory represented absolute truth, in form as well as content. Yet geometry and history have driven us in a particular direction. The parallel postulate could not be proved within Euclid's system, so curved space, though 2,000 years in the queue, was inevitable. Relativity and quantum mechanics were completely independent and philosophically contradictory theories, yet in string theory there seems to exist a third, wildly different theory from which each of those two can be derived. If Hawking's mix of quantum theory and relativity yields a prediction of black hole entropy, and Strominger's unrelated calculation employing string theory agrees, does not this connection imply some deeper truth?

For deeper truths, our search goes on. To Euclid and the geniuses that followed, to Descartes, to Gauss, to Einstein, and—maybe, time will tell—to Witten; and to all those on whose shoulders they stood, we owe a debt of gratitude. They experienced the joy of discovery. For the rest of us they enabled an equal joy, the joy of understanding.

NOTES

2. The Geometry of Taxation

4   Yeats wrote: Yeats referred to Babylonian indifference to knowledge in his poem "The Dawn," which begins:

> I would be ignorant as the dawn
> That has looked down
> On that old queen measuring a town
> With the pin of a brooch,
> Or on the withered men that saw
> From their pedantic Babylon
> The careless planets in their courses,
> The stars fade out where the moon comes,
> And took their tablets and did sums . . .

4   the Ishango bone: Michael R. Williams, *A History of Computing Technology* (Englewood Cliffs, NJ: Prentice-Hall, 1985), pp. 39–40.

4   The thought of performing operations: For a good discussion of the origins of counting and arithmetic, see Williams, chap. 1.

5   the word used for "two": Ibid., p. 3.

5   in the sixth millennium B.C., when the people: R. G. W. Anderson, *The British Museum* (London: British Museum Press, 1997), p. 16.

5   Only the Nile River: Pierre Montet, *Eternal Egypt*, trans. Doreen Weightman (New York: New American Library, 1964), pp. 1–8.

5   "Egypt," means "black earth" in the Coptic language: Alfred Hooper, *Makers of Mathematics* (New York: Random House, 1948), p. 32.

6   Around that time, they also developed writing: Georges Jean, *Writing: The Story of Alphabets and Scripts,* trans. Jenny Oates (New York: Harry N. Abrams, 1992), p. 27.

6   Taxes were perhaps the first imperative: Herodotus wrote that the problem of taxation stimulated the development of Egyptian geometry. See W. K. C. Guthrie, *A History of Greek Philosophy*

267

(Cambridge, UK: University Press, 1971), pp. 34–35, and Herbert Turnbull, *The Great Mathematicians* (New York: New York University Press, 1961), p. 1.

6   100 percent per year: Rosalie David, *Handbook to Life in Ancient Egypt* (New York: Facts on File, 1998), p. 96.

7   with armies who cut the phalluses off: This and other amazing facts can be found in Alexei's contribution to these notes: James Putnam and Jeremy Pemberton, *Amazing Facts About Ancient Egypt* (London and New York: Thames & Hudson, 1995), p. 46.

7   another urbanization occurred: For a good discussion of Babylonian and Sumerian mathematics see Edna E. Kramer, *The Nature and Growth of Modern Mathematics* (Princeton, NJ: Princeton University Press, 1981), pp. 2–12.

7   To the Babylonians we credit: For a comparison of Egyptian and Babylonian mathematics, see Morris Kline, *Mathematical Thought from Ancient to Modern Times* (New York: Oxford University Press, 1972), pp. 11–22. See also H. L. Resnikoff and R. O. Wells, Jr., *Mathematics in Civilization* (New York: Dover Publications, 1973), pp. 69–89.

8   the ruins at Nineveh . . . mostly from the ruins of Nippur and Kis: Resnikoff and Wells, p. 69.

8   Babylonian moneylenders even calculated compound interest: Kline, p. 11.

8   "four is the length . . .": Quoted in *The First Mathematicians* (March 2000), on http://www.members.aol.com/bbyars1/first.html; a similar but more complicated rhetorical problem can be found in Kline, p. 9.

9   the oldest known unambiguous use: Kline, p. 259.

3. Among the Seven Sages

11   The discovery that mathematics is more: Discussion of Thales' life and work can be found in Sir Thomas Heath, *A History of Greek Mathematics* (New York: Dover Publications, 1981), pp. 118–49; Jonathan Barnes, *The Presocratic Philosophers* (London: Routledge & Kegan Paul, 1982), pp 1–16; George Johnston Allman, *Greek Geometry from Thales to Euclid* (Dublin, 1889), pp. 7–17;

G. S. Kirk and J. E. Raven, *The Presocratic Philosophers* (Cambridge, UK: University Press, 1957), pp. 74–98; Hooper, pp. 27–38; and Guthrie, pp. 39–71.

12   for which Miletus was known: Reay Tannahill, *Sex in History* (Scarborough House, 1992), pp. 98–99.

12   archeologists have unearthed a chip of a wine cup: Richard Hibler, *Life and Learning in Ancient Athens* (Lanham, MD: University Press of America, 1988), p. 21.

13   measure the height of the pyramids: Hooper, p. 37.

14   Epicurus still maintained that the sun: Erwin Schroedinger, *Nature and the Greeks* (Cambridge: Cambridge University Press, 1996), p. 81.

14   Thales kept the Egyptian name "earth measurement": Hooper, p. 33.

14   fundamental stuff . . . chose the fish: See Guthrie, pp. 55–80, and Peter Gorman, *Pythagoras, A Life* (London: Routledge & Kegan Paul, 1979), p. 32.

15   living in a city of harbors: To learn about life in Miletus, see Adelaide Dunham, *The History of Miletus* (London: University of London Press), 1915.

15   The historical picture: Gorman, p. 40.

4. The Secret Society

17   Pythagoras took Thales up: The most in-depth biography of Pythagoras, with sources, is Gorman; see also Leslie Ralph, *Pythagoras* (London: Krikos, 1961).

17   Millions of years ago, somebody eeked: See Donald Johanson and Blake Edgar, *From Lucy to Language* (New York: Simon & Schuster, 1996), pp. 106–7.

21   one of Pythagoras' disciples wrote: Jane Muir, *Of Men and Numbers* (New York: Dodd, Mead & Co., 1961), p. 6.

23   attacking a poisonous snake . . . a thief who broke into Pythagoras' home: Gorman, p. 108.

24   Pythagoras, for instance, was believed by many: Ibid., p. 19.

24   he was believed to have the ability: Ibid., p. 110.

24   Both believed in reincarnation: Ibid., p. 111.

24   Pythagoras once stopped a man: Ibid.

25   not urinating in public and not having sex: Ibid., p. 123.

26   the diagonal of a square: For the mathematically inclined, here is the proof. Let c be the length of the diagonal and start by assuming that c can be expressed as a fraction, say, m/n, expressed in lowest terms (i.e., so m and n have no common divisor, in particular that they are not both even). The proof proceeds in three steps. First, note that $c^2 = 2$ means that $m^2 = 2n^2$. In words, that m^2 is an even number. Since the squares of odd numbers are odd, this means that m itself must also be an even number. Second, since m and n cannot both be even, n must be odd. Third, we look at the equation $m^2 = 2n^2$ from another perspective. Since m is even, we can write m as $2q$, for some integer q. If we replace the m in $m^2 = 2n^2$ with $2q$, we get $4q^2 = 2n^2$, which is the same as $2q^2 = n^2$. This means that n^2, hence n, must be even.

     We have just shown that if c can be written as $c = (m/n)$, then n is odd, and n is even. This is a contradiction, so the original assumption, that c can be written $c = (m/n)$, must be false. This type of proof, where we assume the negation of what we wish to prove, and then show that the negation leads to a contradiction, is called *reductio ad absurdum*. It is one of the Pythagorean inventions that remains very useful in mathematics even today.

26   he banned his followers from revealing: Muir, pp. 12–13.

27   Cantor, unable to tolerate this, had a breakdown: Kramer, p. 577.

27   The Romans hated the long hair: Gorman, pp. 192–93.

5. Euclid's Manifesto

29   Spinoza emulated . . . Kant defended him: Spinoza, an important seventeenth-century philosopher, wrote his chief work, the *Ethics,* in the style of Euclid's elements, starting with definitions and axioms, from which he claimed to prove rigorous theorems. *Ethics* is available on the web at the Middle Tennessee State University website: Baruch Spinoza, *Ethics,* trans. by R. H. M. Elwes (1883), MTSU Philosophy WebWorks Hypertext Edition (1997), http://www.frank.mtsu.edu/~rbombard/RB/spinoza/ethica-front.html. See also Bertrand Russell, *A History of Western Philosophy* (New York: Simon & Schuster, 1945), p. 572. Abraham Lincoln, while still an obscure lawyer, studied *Elements* to improve his logical

NOTES

skills—see Hooper, p. 44. Kant believed that Euclidean geometry is hard-wired in the human brain—see Russell, p. 714.

29   All we know is: Heath, pp. 354–55.

29   later momentous work by Apollonius: Kline, pp. 89–99, 157–58.

29   The *Elements* has a history: Heath, pp. 356–70; see also Hooper, pp. 44–48. In 1926, Heath personally added to the history of *Elements* by releasing his own edition, reprinted by Dover: Sir Thomas Heath, *The Thirteen Books of Euclid's Elements* (New York: Dover Publications, 1956).

30   a theorem of logic states: Kline, p. 1205.

31   Bayes' theorem: The *Let's Make a Deal* quandary is usually called the Monty Hall problem, after the show's host, Monty Hall. The best way to understand the resolution is to draw a tree diagram illustrating the successive possible choices. This method is used to illustrate Bayes' theorem in John Freund, *Mathematical Statistics* (Englewood, Cliffs, NJ: Prentice-Hall, 1971), pp. 57–63.

33   A trick invented by Paul Curry: Martin Gardner, *Entertaining Mathematical Puzzles* (New York: Dover Publications, 1961), p. 43.

33   a deviation from classical Newtonian theory: For the history of the perihelion problem, see John Earman, Michael Janssen, and John D. Norton, eds., *The Attraction of Gravitation: New Studies in the History of General Relativity* (Boston: The Center for Einstein Studies, 1993), pp. 129–49. There is also a good, but brief, discussion in Abraham Pais, *Subtle Is the Lord* (Oxford: Oxford University Press, 1982), pp. 22, 253–55; the Leverrier quote is given on p. 254; the "high point" on p. 22. There is also a good discussion of the geometry of the situation in Resnikoff and Wells, pp. 334–36.

34   He stated twenty-three definitions: Euclid's *Elements*, with some commentary, can be found in Heath, pp. 354–421. Three good, and more modern, discussions appear in Kline, *Mathematical Thought*, pp. 56–88; Jeremy Gray, *Ideas of Space* (Oxford: Clarendon Press, 1989), pp. 26–41; and Marvin Greenberg, *Euclidean and Non-Euclidean Geometries* (San Francisco: W. H. Freeman & Co., 1974), pp. 1–113.

35   They were non-geometric assertions: Kline, p. 59.

6. A Beautiful Woman, a Library, and the End of Civilization

39   The Macedonians: H. G. Wells, *The Outline of History* (New York: Garden City Books, 1949), pp. 345–75. For a timeline, see Jerome Burne, ed., *Chronicle of the World* (London: Longman Chronicle, 1989), pp. 144–47.

39   he ordered leading Macedonian citizens: Russell, p. 220.

40   Ptolemy III wrote . . . then kept them: The Athenians lent Ptolemy III treasured manuscripts of Euripedes, Aeschylus, and Sophocles. Though he kept them, Ptolemy III was generous enough to return copies he had made. The Greeks must not have been totally surprised. They had asked for (and kept) a fortune in collateral which he had been required to offer. See Will Durant, *The Life of Greece* (New York: Simon & Schuster, 1966), p. 601.

41   Eratosthenes of Cyrene: The geometry of his calculation is explained in Morris Kline, *Mathematics and the Physical World* (New York: Dover Publications, 1981), pp. 6–7.

41   a stick in the ground casts no shadow: There are differing accounts of the story. In some, Eratosthenes notices the lack of shadows by gazing down a well, and determines the distance to Syene employing travelers' reports. The version quoted here can be found in Carl Sagan, *Cosmos* (New York: Ballantine Books, 1981), pp. 6–7.

42   we do know who discovered the principle: Kline, *Mathematical Thought*, p. 106.

43   Archimedes was so proud: Morris Kline, *Mathematics in Western Culture* (London: Oxford University Press, 1953), p. 66.

43   Astronomy, too, reached an apex: Kline, *Mathematical Thought*, pp. 158–59.

43   Ptolemy also wrote a book called *Geographia:* For a summary of Ptolemy's work, see John Noble Wilford, *The Mapmakers* (New York: Vintage Books, 1981), pp. 25–33.

45   In a Roman textbook: Kline, *Mathematics in Western Culture*, p. 86.

45   Anicius Manlius Severinus Boethius: Kline, *Mathematical Thought*, p. 201.

46   *Topographia Christiana:* Kline, *Mathematics in Western Culture*, p. 89.

46   Hypatia, the first great woman scholar: For Hypatia's story, see

Maria Dzielska, *Hypatia of Alexandria,* trans. F. Lyra (Cambridge, MA: Harvard University Press, 1995). See also Kramer, pp. 61–65, and Russell, pp. 367–69.

47 *The Decline and Fall of the Roman Empire:* Edward Gibbon, *The Decline and Fall of the Roman Empire* (London: 1898), pp. 109–10.

47 On October 15, 412: Dzielska, p. 84.

47 According to Damascius: Ibid., p. 90.

48 There are several versions: Ibid., pp. 93–94.

48 A recent historical study estimates: Resnikoff and Wells, pp. 4–13.

49 By 800, only fragments: David Lindberg, ed., *Science in the Middle Ages* (Chicago: University of Chicago Press, 1978), p. 149.

7. The Revolution in Place

58 as one old textbook on the subject states: William Gondin, *Advanced Algebra and Calculus Made Simple* (New York: Doubleday & Co., 1959), p. 11.

8. The Origin of Latitude and Longitude

55 some of the earliest known maps: Two excellent accounts of the history of mapmaking are Wilford; and Norman Thrower, *Maps and Civilization* (Chicago: University of Chicago Press, 1996).

56 Polaris hasn't always been: Resnikoff and Wells, pp. 86–89.

58 an error of just three seconds: Dava Sobel, *Longitude* (New York: Penguin Books, 1995), p. 59.

58 Finally, in October 1884: Wilford, pp. 220–21.

9. The Legacy of The Rotten Romans

60 Charlemagne: Morris Bishop, *The Middle Ages* (Boston: Houghton Mifflin, 1987), pp. 22–30.

61 Carolingian miniscule: Jean, pp. 86–87.

63 Bartholomew wrote: Jean Gimpel, *The Medieval Machine* (New York: Penguin Books, 1976), p. 182.

63   The medieval mathematician faced: Bishop, pp. 194–95.

64   Europe at the time: Robert S. Gottfried, *The Black Death* (New York: The Free Press, 1983), pp. 24–29.

64   The Florentine historian Giovanni Villani: Ibid., p. 53.

64   College provided no haven: For a description of the medieval university and university life, see Bishop, pp. 240–44, and Mildred Prica Bjerken, *Medieval Paris* (Metuchen, NJ: Scarecrow Press, 1973), pp. 59–73.

65   The science of the day: Bishop, pp.145–46.

65   Frederick indulged his love: Ibid., pp., 70–71.

65   The concept of time: Gimpel, pp. 147–70; Bishop, pp. 133–34.

66   Cartography, too, was primitive: Wilford, pp. 41–48; Thrower, pp. 40–45.

67   the Scholastics: Russell, pp. 463–75. For Abelard, see also Jacques LeGoff, *Intellectuals in the Middle Ages,* trans. Teresa Lavender Fagan (Oxford: Blackwell, 1993), pp. 35–41.

68   from the village of Allemagne: Jeannine Quillet, *Autour de Nicole Oresme* (Paris: Librairie Philosophique J. Vrin, 1990), pp. 10–15.

10. The Discreet Charm of the Graph

70   a fourteenth-century English recipe: Reay Tannahill, *Food in History* (New York: Stein & Day, 1973), p. 281.

72   a whole new discipline: The theory of distributions. For the mathematically inclined, an excellent and classic reference on the undergraduate level is M. J. Lighthill, *Introduction to Fourier Analysis and Generalised Functions* (Cambridge, UK: University Press, 1958).

72   Nicole d'Oresme: For Oresme's work on the graph, see Lindberg, pp. 237–41; Marshall Clagett, *Studies in Medieval Physics and Mathematics* (London: Variorum Reprints, 1979), pp. 286–95; Stephano Caroti, ed., *Studies in Medieval Philosophy* (Leo S. Olschki, 1989), pp. 230–34.

76   the Merton rule: David C. Lindberg, *The Beginnings of Western Science* (Chicago: University of Chicago Press, 1992), pp. 290–301.

77   Oresme also applied his graphical reasoning: Clagett, pp. 291–93.

77   Another scoop he got on Galileo: Lindberg, *The Beginnings,* pp. 258–61.

78   His conversion came: Ibid., pp. 260–61.

78   Oresme wrote, in the tradition of Socrates: Charles Gillespie, ed.,
      The Dictionary of Scientific Biography (New York: Charles Scrib-
      ner's Sons, 1970–1990).

11. A Soldier's Story

79   On March 31, 1596: The best modern biography of Descartes is
      Jack Vrooman, *René Descartes* (New York: G. P. Putnam's Sons,
      1970). For accounts of his life interweaving his mathematics, see
      Muir, pp. 47–76; Stuart Hollingdale, *Makers of Mathematics* (New
      York: Penguin Books, 1989), pp. 124–36; Kramer, pp. 134–66;
      and Bryan Morgan, *Men and Discoveries in Mathematics* (Lon-
      don: John Murray, 1972), pp. 91–104.

79   some say ten: Various references differ on this point. It seems to be
      evenly split.

81   Descartes later described Beekman: Muir, p. 50.

81   Descartes wrote that "it already wearies me . . .": George Molland,
      Mathematics and the Medieval Ancestry of Physics (Aldershot,
      Hampshire, U.K., and Brookfield, VT: 1995), p. 40.

81   Descartes wrote "only on the condition . . .": Kline, *Mathematical
      Thought,* p. 308.

81   scholars write that "Descartes's mathematical laziness . . .": Mol-
      land, p. 40.

83   the use of coordinates alone: For a description of Ptolemy's work,
      see Wilford, pp. 25–34. A few decades before Descartes was born,
      in 1569, mapmaking had its own revolution when Gerhard Kremer,
      better known by his latinized name Gerardus Mercator, published a
      new kind of world map. With this map, Mercator solved the prob-
      lem of projecting the sphere of the earth onto a flat map, and he did
      it in a way particularly useful to navigators. Although Mercator's
      map stretched and contracted distances, the angles between curves
      on his map remained true. That is, they were the same on his flat
      map as on the curved earth. This was significant because the easi-
      est course for a ship's helmsman to follow was a course in which
      he kept a fixed angle with north as shown by the compass needle.
      Mathematically, the map was significant because it represented a
      manipulation, or transformation, of coordinates. Mercator himself

did none of this mathematics—he derived his map empirically. Cartesian geometry allows the analysis to be carried out mathematically, resulting in a much greater insight into mapmaking. Descartes knew of Mercator's map, but we don't know if, or how much, Descartes was influenced by advances in the science of cartography because he neglected to put citations in any of his publications. For a discussion of the mathematics behind Mercator's work, see Resnikoff and Wells, pp. 155–68.

84   in the absence of better notation: Descartes did not simply inherit all the algebra he needed for his work. He invented much of it himself. First, he invented the modern notation of employing the last letters of the alphabet to represent unknown variables, and the first letters to represent constants. Prior to Descartes, the language of algebra was rather awkward. For instance, what Descartes wrote as $2x^2 + x^3$ would have been written in words, "2Q plus C," where Q stood for the square (*carré*) and C for the cube. Descartes's notation is superior in that it makes explicit both the unknown quantity you are squaring and cubing (x), and the nature of the powers of x (2 and 3) that are taken. Employing this more elegant notation, Descartes could add or subtract equations, or perform other arithmetic operations on them. He was able to classify algebraic expressions according to the type of curve they represented. For instance, he would recognize the equations $3x + 6y - 4 = 0$ and $4x + 7y + 1 = 0$ as both representing lines, which he studied in the more general case $ax + by + c = 0$. In this way, he transformed algebra from the study of a hodgepodge of individual equations to the study of whole classes of equations—see Vrooman, pp. 117–18. For a more general history of algebraic symbolism, see Kline, *Mathematical Thought,* pp. 259–63, and Resnikoff and Wells, pp. 203–6.

84   a table of the approximate average high temperatures: As read from a table published in the *New York Times,* January 11, 1981, and reproduced in Tufte.

86   That means that the square of the distance: We can now understand better Descartes's definition of a circle. If the circle is centered at the origin of coordinates, and the coordinates of a point along its circumference are x and y, then demanding that x and y satisfy the equation $x^2 + y^2 = r^2$ simply means that we require that all points

along the circumference be a distance r from the center, the simple intuitive definition we know from our school days.

86   Descartes's formula for distance: Though we have discussed the plane, a two-dimensional space, Descartes's coordinates are easy to extend to three or more dimensions. For instance, the equation for a sphere is $x^2 + y^2 + z^2 = r^2$: the only change is the addition of a new coordinate, z. In this way, physical theories can actually be written for an arbitrary number of spatial dimensions. It turns out, for instance, that ordinary quantum mechanics takes an especially simple form in an infinite number of space dimensions, and this property has been used to approximate the solutions of equations that are otherwise hard to solve. The mathematically inclined can find this in L. D. Mlodinow and N. Papanicolaou, "SO(2,1) Algebra and Large N Expansions in Quantum Mechanics," *Annals of Physics,* vol. 128, no. 2 (September, 1980), pp. 314–34.

86   "my entire physics is nothing other than geometry": Vrooman, p. 120.

87   "It seems to me that he [Galileo] lacks . . .": Ibid., p. 115.

87   he canceled publication: Ibid., pp. 84–85.

87   the original manuscript had the snappy title: Ibid., p. 89.

88   Descartes was viciously attacked: Ibid., pp. 152–55; 157–62.

88   one affair in his life: Ibid., pp. 136–49.

12. Iced by the Snow Queen

90   Queen Christina: For an account of Descartes and Christina, see Vrooman, pp. 212–55.

92   except for his skull: For the story of the travels of Descartes's different body parts after death, see ibid., pp. 252–54.

13. The Curved Space Revolution

96   It was one of the first books: Heath, pp. 364–65.

14. The Trouble with Ptolemy

98   The first known attempt: For Ptolemy's and Proclus' arguments, see Kline, *Mathematical Thought*, pp. 863–65.

102   Baghdad scholar Thābit ibn Qurrah: The Islamic civilization in medieval times contributed greatly to the development of mathematics, not only in preserving Greek works but in the development of algebra. One good account of this is J. L. Berggren, *Episodes in the Mathematics of Medieval Islam* (New York: Springer-Verlag, 1986); a brief account of Thābit ibn Qurrah's life can be found on pp. 2–4. His attempt at proving the parallel postulate is described in Gray, pp. 43–44. Attempts by later Islamic mathematicians are also described in Gray.

104   the parallel postulate is now easy to prove: For details, see Gray, pp. 57–58.

15. A Napoleonic Hero

107   In Göttingen, on February 23, 1855: For a detailed account of Gauss's life, see G. Waldo Dunnington, *Carl Friedrich Gauss: Titan of Science* (New York: Hafner Publishing Co., 1955).

107   For most of his life: Muir, p. 179.

107   "a burdensome, ungratifying business": Ibid., p. 181.

107   "without immortality, the world would be meaningless": Ibid., p. 182.

107   "the griefs overbalance the joys a hundredfold": Ibid., p. 179.

109   calling him "domineering, uncouth, and unrefined": Ibid., p. 161.

111   "one is only moderately prepared . . .": Hollingdale, p. 317.

113   "For three days, that angel . . .": Ibid., p. 65.

114   "It is true that in my life . . .": Muir, p. 179.

16. The Fall of the Fifth Postulate

115   Gauss would dismiss Kaestner: Dunnington, p. 24.

116   Gauss wrote to F. A. Taurinus: Ibid., p. 181.

117   Hobbes and Wallis: Russell, p. 548; for details, see http://www.turnbull.dcs.st-and.ac.uk/history/Mathematicians/Wallis.html (from the St. Andrews College website, April 99).

117 The philosopher whose followers: Kline, *Mathematical Thought,* p. 871.

117 the pretense of rigor: Russell, *Introduction to Mathematical Philosophy* (New York: Dover Publications 1993), pp. 144–45.

117 Gauss held the opposite view: Dunnington, p. 215.

117 Kant calls Euclidean space: See Greenberg, p. 146. For two good analyses of Kant's views on space and time, see Russell, *Introduction to Mathematical Philosophy,* pp. 712–18, and Max Jammer, *Concepts of Space* (New York: Dover Publications, 1993), pp. 131–39.

117 Χωριάτικη Σαλάτα: A typical Greek salad.

118 the distinction between analytic and synthetic judgments: from *Critique of Pure Reason,* Vol. IV.

118 Richard Feynman, when asked: I had many personal discussions with Feynman on this point at the California Institute of Technology, in Pasadena, in the period 1980–82.

118 had "created a new, different world . . .": Dunnington, p. 183. For more details on the life and work of Bolyai, see Gillespie, *Dictionary of Scientific Biography,* pp. 268–71. For Lobachevsky, see Muir, pp. 184–201; E. T. Bell, *Men of Mathematics* (New York: Simon & Schuster, 1965), pp. 294–306; and Heinz Junge, ed., *Biographien bedeutender Mathematiker* (Berlin: Volk und Wissen Volkseigener Verlag, 1975), pp. 353–66.

119 at least one song about Lobachevsky: "Nicolai Ivanovitch Lobachevoki" by Tom Lehrer. As of publication, the text was available at http://www.keaveny.demon.co.uk/lehrer/lyrics/maths.htm

119 János Bolyai never published another work: Strangely, papers found after Bolyai's death revealed that he was a closet Euclidean: even after his discovery of non-Euclidean space, he continued to attempt to prove the Euclidean form of the parallel postulate, which would have debunked his own work.

120 He took pleasure in the collection: Dunnington, p. 228.

17. Lost in Hyperbolic Space

121 "Mathematicians are born, not made": "Quotations by Henri Poincare" in http://www-groups.dcs.st-and.ac.uk/history/Mathemati

cians/Quotations/Poincare.html (from the St. Andrews College website, June, 1999).

122 a concrete model of hyperbolic space: For a detailed mathematical discussion of the Poincaré model, see Greenberg, pp. 190–214.

122 "any arcs of circles . . .": To be mathematically correct, we should note that there is also another type of curve called a line in Poincaré's model. It is a diameter, that is, any line segment that passes through the central point of the crèpe and has endpoints on the crèpe's boundary. This isn't really different from the other kinds of Poincaré-lines: a diameter is perpendicular to the boundary of the crépe, and it can be considered to be an arc of a circle—an infinitely large circle.

126 elliptic spaces cannot exist: In the early eighteenth century Gerolamo Saccheri, a Jesuit priest and professor at the University of Pavia, studied the work of Thābit's follower Nāsir-Eddin and Wallis. Inspired by them, he set out as others had to vindicate Euclid of all faults. We know this was his motive because in the year of his death, 1733, Saccheri published a book called *Euclid Vindicated from All Faults (Euclides ab Omni Maevo Vindicatus)*. Like those before him, Saccheri was wrong. But he did correctly prove one thing: that the form of the parallel postulate that leads to elliptic space also leads to logical contradictions with Euclid's other axioms.

18. Some Insects Called the Human Race

127 In the decade starting with 1816: For a history of Gauss's work on geodesy, see Dunnington, pp. 118–38.

133 certain genius birds . . . : Interview with Steven Mlodinow, October 9, 1999.

19. A Tale of Two Aliens

136 Georg Riemann: An excellent account of Riemann's work and legacy, with some biographical content, can be found in Michael Monastyrsky, *Riemann, Topology, and Physics,* trans. Roger Cooke, James King, and Victoria King (Boston: Birkhauser, 1999).

A summary of Riemann's life can also be found in Bell, pp. 484–509.

136 Adrien-Marie Legendre's book: 2 vols., 1830 (Paris: A. Blanchard, 1955). For the story of Riemann's quick reading of it, see Bell, p. 487.

137 Gauss wrote that Riemann: Bell, p. 495.

141 "In a complete development . . .": Quoted in Kline, *Mathematical Thought,* p. 1006.

20. After 2,000 Years, A Face-lift

144 To paraphrase the great Göttingen mathematician: David Hilbert, *Grundlagen der Geometrie* (Berlin: B. G. Teubner, 1930). This quote is discussed in Kline, *Mathematical Thought,* pp. 1010–15, and Greenberg, pp. 58–59. Greenberg also has a good discussion of undefined terms on pp. 9–12.

145 In 1871, the Prussian mathematician Felix Klein: Gray, p. 155.

146 In 1894, Italian logician Giuseppe Peano: Kline, *Mathematical Thought,* p. 1010.

146 In 1899, Hilbert: For a more in-depth presentation of Hilbert's axioms, see Greenberg, pp. 58–84.

146 In Hilbert's system: Kline, *Mathematical Thought,* pp. 1010–15.

149 the shocking theorem of Kurt Gödel: For an excellent explication, see Ernest Nagel and James R. Newman, *Gödel's Proof* (New York: New York University Press, 1958), and the more wide-ranging classic it inspired, Douglas Hofstadter, *Godel, Escher, Bach: An Eternal Golden Braid* (New York: Vintage Books, 1979).

21. Revolution at the Speed of Light

153 "The question of the validity . . .": Monastyrsky, p. 34.

153 Clifford boldly proclaimed: Ibid., p. 36.

154 some say of exhaustion: For example, J. J. O'Connor and E. F. Robertson, *William Kingdon Clifford,* http://www-groups.dcs.st-and.ac.uk/history/Mathematicians/Clifford.html (From the St. Andrews College website, June 1999).

22. Relativity's Other Albert

157   Albert's family, the Michelsons: For the story of Michelson's life, see Dorothy Michelson Livingston, *The Master of Light: A Biography of Albert A. Michelson* (New York: Scribner, 1973).

158   Michelson eventually got to see: See Harvey B. Lemon, "Albert Abraham Michelson: The Man and the Man of Science," *American Physics Teacher* (now *American Journal of Physics*), vol. 4, no. 2 (February 1936).

158   Grant had excelled in mathematics: Brooks D. Simpson, *Ulysses S. Grant: Triumph Over Adversity 1822–1865* (New York: Houghton Mifflin, 2000), p. 9.

159   "If you'd give less attention to scientific things . . .": *New York Times,* May 10, 1931, p. 3, cited in Daniel Kevles, *The Physicists* (Cambridge, MA: Harvard University Press, 1995), p. 28.

159   ". . . there is a subtle, imponderable . . .": Adolphe Ganot, *Eléments de Physique,* ca.1860, quoted in Loyd S. Swenson, Jr., *The Ethereal Aether.* (Austin, TX: University of Texas Press, 1972), p. 37.

160   The modern concept of the ether: G. L. De Haas-Lorentz, ed., *H. A. Lorentz* (Amsterdam: North-Holland Publishing Co., 1957), pp. 48–49.

160   The term was Aristotle's: For a discussion of Aristotle's ether, see Henning Genz, *Nothingness: The Science of Empty Space* (Reading, MA: Perseus Books, 1999), pp. 72–80.

160   "One knows where our belief . . .": Pais, p. 127.

161   opined E. S. Fischer: The passage reads, "We do not know what this medium is, and we seem destined to remain ignorant since we cannot perceive the medium itself, but only the objects which become visible by its influence. . . . Meanwhile it is of no consequence to us . . . provided we know the laws of the phenomena; and these laws have actually been developed almost as perfectly as those of gravity."—E. S. Fischer, *Elements of Natural Philosophy* (Boston, 1827), p. 226. The English edition was translated from German into French by M. Biot, the famous thermodynamicist, and then from French into English.

161   Fresnel showed that light waves: He was actually reacting to the discovery of polarized light in 1808 by the French physicist Etienne-Louis Malus. According to Fresnel, polarization was possible because light could vibrate in either of the two directions per-

pendicular to its path. Filtering out one or the other of these is what leads to polarization. Waves that vibrate only along their direction of motion cannot have this property.

23. The Stuff of Space

163 The author's birth name was James Clerk: Two biographies of Maxwell, written about 100 years apart, are Louis Campbell and William Garnet, *The Life of James Clerk Maxwell* (London, 1882; New York: Johnson Reprint Co., 1969), and Martin Goldman, *The Demon in the Aether* (Edinburgh: Paul Harris Publishing, 1983).

164 The equations were Maxwell's equations: For the mathematically inclined, Maxwell's equations in free space are:
$$\nabla \cdot \mathbf{E} = 4\,\pi\,\rho;\ \nabla \cdot \mathbf{B} = 0;\ \nabla \times \mathbf{B} - \partial\mathbf{E}/\partial t = 4\,\pi\mathbf{j};\ \nabla \times \mathbf{E} + \partial\mathbf{B}/\partial t = 0,$$
where ρ and $\mathbf{j}$ are the sources and $\mathbf{E}$ and $\mathbf{B}$ are the fields.

165 "It is not always easy to comprehend . . .": Haas-Lorentz, ed., p. 55.

165 "a kind of intellectual jungle": Ibid., p. 55.

165 "Whatever difficulties we may have . . .": James Clerk Maxwell, "Ether," *Encyclopaedia Britannica,* 9th edn., Vol. VIII (1893), p. 572, quoted in Swenson, p. 57.

169 the advances in precision tooling: Swenson, p. 60.

169 Fizeau's measurement came within: Ibid., pp. 60–62.

172 Sir William Thomson (Lord Kelvin), on a visit: In a lecture delivered in Philadelphia at the Academy of Music, September 24, 1884. A transcript of the talk appears as Sir William Thomson [Lord Kelvin], "The Wave Theory of Light," in Charles W. Elliot, ed., *The Harvard Classics,* Vol. 30, *Scientific Papers,* p. 268. It is quoted in Swenson, p. 77.

172 "I have repeatedly tried . . .": Swenson, p. 88.

172 he questioned Michelson's theoretical analysis: Ibid., p. 73.

173 they too, lost interest: Michelson would later repeat his experiment several times throughout his career, as did others, most notably his successor when he left Case, Dayton Clarence Miller. Michelson would never come to accept that the ether did not exist. As late as 1919, Einstein hoped to get Michelson's support for his theory. The closest he came was an equivocal paragraph in a book Michelson published in 1927, a few years before his death. See Denis

Brian, *Einstein, A Life* (New York: John Wiley & Sons, 1996), pp. 104, 126–27, 211–13, and Pais, pp. 111–15.

173 "I have read with much interest . . .": G. F. FitzGerald, *Science*, vol. 13 (1889), p. 390, quoted in Pais, p. 122.

174 tried to construct an explanation: Kenneth F. Schaffner, *Nineteenth-Century Aether Theories* (Oxford: Pergamon Press, 1972), pp. 99–117.

175 "There is no absolute time . . ." Poincaré's remarks were published in a book, *La Science et l'Hypothèse*, and pored over by Einstein and some of his friends in Bern. The book was reprinted as Henri Poincaré, *Science and Hypothesis* (New York: Dover Publications, 1952).

24. Probationary Technical Expert, Third Class

176 Einstein was no boy genius: There are many biographies of Einstein. Two I found especially useful were the one by Brian; and Ronald Clark, *Einstein: The Life and Times* (London: Hodder & Stoughton, 1973; New York: Avon Books, 1984). In addition, Pais is an excellent scientific biography which has the advantage of a personal perspective.

178 *hocked a chainik:* Literally, "beat a kettle"; the expression means roughly "talked his ear off."

179 "One had to cram . . .": Quoted in Hollingdale, p. 373.

180 "A New Determination of Molecular Dimensions": "Eine neue Bestimmung der Moleküldimensionen," *Annalen der Physik*, vol. 19 (1906), p. 289.

180 According to a study: Pais, pp. 89–90.

25. A Relatively Euclidean Approach

182 "On the Electrodynamics of Moving Bodies": *Annalen der Physik*, vol. 17 (1905), p. 891. An English translation appears in A. Somerfeld, *The Principle of Relativity* (New York: Dover Publications, 1961), p. 37.

182 "Here were assertions . . .": Hollingdale, p. 370.

184 in his 1916 book, *Relativity:* Albert Einstein, *Relativity*, trans. Robert Lawson (New York: Crown Publishers, 1961).

188 defined now (for technical reasons): In relativity, time is considered a dimension, but in flat or nearly flat space-time the separation, the relativistic version of distance is defined in terms of time differences *minus* spatial differences. This means, for instance, that the shortest path between two events with zero time difference is the path (a line through space) with the largest (i.e., least negative) separation.

190 a "venerable federal ink shitter": See Brian, p. 69.

191 When Einstein entered the room: See ibid., pp. 69–70.

192 In a 1908 lecture, Minkowski said: Quoted in Pais, p. 152. Unfortunately, a few months later, Minkowski died suddenly of appendicitis.

192 "form a modest crowd": Pais, p. 151.

192 Lorentz, who shared: Pais, pp. 166–7.

192 Poincaré, who never understood: Pais, pp. 167–71.

26. Einstein's Apple

193 "I was sitting in a chair . . .": Pais, p. 179.

193 "the happiest thought of my life": Ibid., p. 178.

199 "It is impossible to distinguish . . .": For this formulation of the equivalence principle, see Charles Misner, Kip Thorne, and John Wheeler, *Gravitation* (San Francisco: W. H. Freeman & Co., 1973), p. 189.

202 an extra minute a year: Ibid., p. 131.

202 gravitational redshift: The effect was observed in 1960 by R. V. Pound and G. A. Rebka, Jr., *Physical Review Letters,* vol. 4 (1960), p. 337.

203 "If we knew what . . .": http://stripe.colorado.edu/~judy/einstein/science.html (June, 1999).

203 "Because of the Lorentz contraction . . .": Pais, p. 213.

27. From Inspiration to Perspiration

205 "Grossmann, you must help me . . .": Pais, p. 212.

205 "a terrible mess which physicists . . .": Ibid., p. 213.

205 "in all my life I have labored . . .": Ibid., p. 216.

206 "As an older friend . . .": Ibid., p. 239.

206　On November 25, 1915: Five days earlier, on November 20, Hilbert had presented a derivation of the same equations to the Royal Academy of Sciences in Göttingen. His derivation was independent of Einstein's, and in some ways superior, but this was only the last step in the theory, which Hilbert recognized was Einstein's creation. Einstein and Hilbert admired each other and never quarreled over priority. As Hilbert said, "Einstein did the work and not the mathematicians." See Jagdish Mehra, *Einstein, Hilbert, and the Theory of Gravitation* (Boston: D. Reidel Publishing Co., 1974), p. 25.

206　"Finally the general theory . . .": Pais, p. 239.

207　Mathematically, this reads: Actually, except when employing rectangular coordinates in flat space-time, this definition applies to infinitesimal regions only, and then the distances must be added up employing calculus. Mathematically, we write: $ds^2 = g_{11}dx_1^2 + g_{12}dx_1dx_2 + \ldots + g_{34}dx_3dx_4 + g_{44}dx_4^2$.

207　(in four dimensions the metric has ten independent components): The ten components are g_{11}, g_{12}, g_{13}, g_{14}, g_{22}, g_{23}, g_{24}, g_{33}, g_{34}, and g_{44}, where we have eliminated redundancy by applying $g_{ij} = g_{ji}$.

208　For the sun, it is half a kilometer: See Richard Feynman, Robert Leighton, and Matthew Sands, *The Feynman Lectures on Physics,* Vol. II (Reading, MA: Addison-Wesley, 1964), chap. 42, pp. 6–7.

208　Global Positioning Satellites: Marcia Bartusiak, "Catch a Gravity Wave," *Astronomy,* October, 2000.

28. Blue Hair Triumphs

210　the one that was to be successful: Some scientists now feel Eddington may have fudged some of his results. See, for example, James Glanz, "New Tactic in Physics: Hiding the Answer," *New York Times,* August 8, 2000, p. F1.

210　"The present eclipse expeditions . . .": Pais, p. 304.

210　the result was announced: For a description of Eddington's expedition and the reaction, see Clark, pp. 99–102.

211　"The Einstein theory is a fallacy . . .": Brian, pp. 102–3.

212　"The most important example . . .": Ibid., p. 246.

212　In 1931, a booklet entitled: See "The Reaction to Relativity Theory in Germany III: 'A Hundred Authors Against Einstein,'" in John

Earman, Michel Janssen, and John Norton, eds., *The Attraction of Gravitation* (Boston: Center for Einstein Studies, 1993), pp. 248–73.

212 ". . . unfortunately, his [Einstein's] friends . . .": Brian, p. 284.

213 the deciding factor: Brian, p. 233.

213 "What God has torn asunder . . ." Brian, p. 433.

213 "I am generally regarded . . .": Pais, p. 462.

214 "I do not believe . . .": Ibid., p. 426.

214 "When a blind beetle . . .": http://stripe.colorado.edu/~judy/ein stein/himself.html (April, 1999).

29. The Weird Revolution

218 string theory should ultimately be: Ivars Peerson, "Knot Physics," *Science News*, vol. 135, no. 11, March 18, 1989, p. 174.

30. Ten Things I Hate About Your Theory

220 future storm trooper Pascual Jordan: Engelbert L. Schucking, "Jordan, Pauli, Politics, Brecht, and a Variable Gravitational Constant," *Physics Today* (October 1999), pp. 26–31.

221 Murray Gell-Mann describes this way: Interview with Murray Gell-Mann, May 23, 2000.

221 He once wrote, "It has never happened . . .": Walter Moore, *A Life of Erwin Schroedinger* (Cambridge, UK: University Press, 1994), p. 195.

221 what Princeton mathematician Hermann Weyl: Moore, p. 138.

31. The Necessary Uncertainty of Being

223 "The theory [quantum mechanics] yields much . . .": Einstein quote from a letter to Max Born, December 4, 1926, Einstein Archive 8–180; quoted in Alice Calaprice, ed., *The Quotable Einstein* (Princeton, NJ: Princeton University Press, 1996).

225 In 1964, the American physicist John Bell: Bell published his proposal in a short-lived journal called *Physics*. The usual experimental verification cited by physicists is A. Aspect, P. Grangier, and

G. Roger, *Physical Review Letters*, vol. 49 (1982). A later refinement can be found in Gregor Weihs et al., *Physical Review Letters*, vol. 81 (1998).

32. Clash of the Titans

230 "the best tested theory on earth . . .": Toichiro Kinoshita, "The Fine Structure Constant," *Reports on Progress in Physics*, vol. 59 (1996), p. 1459.

33. A Message in a Kaluza-Klein Bottle

231 Einstein wrote back, "The idea . . .": Pais, p. 330.
232 Einstein wrote, "The formal unity . . .": Ibid.
234 In 1926, Einstein described the conditions: *Dictionary of Scientific Biography*, pp. 211–12.

34. The Birth of Strings

235 Gabriele Veneziano: Interview with Gabriele Veneziano, April 10, 2000.

35. Particles, Schmarticles!

239 he believed that the universe: George Johnson, *Strange Beauty* (New York: Alfred A. Knopf, 1999), pp. 195–96.
239 Witten calls S-matrix theory: Interview with Ed Witten, May 15, 2000.
239 Gell-Mann says it was overblown: Interview with Murray Gell-Mann, May 23, 2000.
240 J. Robert Oppenheimer suggested: Quoted in Michio Kaku, *Introduction to Superstrings and M-Theory* (New York: Springer-Verlag, 1999), p. 8.
240 Enrico Fermi remarked: Quoted in Nigel Calder, *The Key to the Universe* (New York: Penguin Books, 1977), p. 69.

242   (the Fermi coupling constant): Constants taken from P. J. Mohr and B. N. Taylor, "CODATA Recommended Values of the Fundamental Constants: 1998," *Review of Modern Physics,* vol. 72 (2000).

243   This fundamental frequency: For a good explanation of the music of strings, see Kline, *Mathematics and the Physical World,* pp. 308–12; and, for more depth, Juan Roederer, *Introduction to the Physics and Psychophysics of Music,* 2nd edn. (New York: Springer-Verlag, 1979), pp. 98–119.

247   They are called Calabi-Yau spaces: P. Candelas et al., *Nuclear Physics,* B258 (1985), p. 46.

247   but those are technical details: Technically, by having holes, physicists mean having the appropriate value of a mathematical quantity called the Euler characteristic (or number), which can be calculated for each Calabi-Yau space. The Euler characteristic is a topological concept that in two or three dimensions is easily visualized, but can also be applied to higher dimensions. In three dimensions, a solid object has an Euler characteristic of two, be it a cube, a sphere, or a soup bowl, whereas objects with holes or handles, like a donut, a coffee cup, or a beer mug, have an Euler characteristic of zero.

36. The Trouble with Strings

250   Gell-Mann, who was working: Quotes in this paragraph are from an interview with Murray Gell-Mann, May 23, 2000.

252   "People didn't want to make the investment": Interview with John Schwarz, March 30, 2000.

252   The few papers he did with Schwarz: Ibid.

252   "I couldn't get John a regular": Interview with Murray Gell-Mann, May 23, 2000.

253   says Schwarz, "it is not clear": Interview with John Schwarz, July 13, 2000.

253   "It made me happy and proud": Interview with Murray Gell-Mann, May 23, 2000.

254   One administrator commented: Ibid.

254   Witten says, "Without John Schwarz . . .": Interview with Ed Witten, May 15, 2000.

37. The Theory Formerly Known As Strings

255   the *Los Angeles Times* had gone as far: Quoted in K. C. Cole, "How Faith in the Fringe Paid Off for One Scientist," *L.A. Times,* November 17, 1999, p. A1.

255   Lamented string theorist Andrew Strominger: Faye Flam, "The Quest for a Theory of Everything Hits Some Snags," *Science,* June 6, 1992, p. 1518.

255   To paraphrase Strominger: Strominger quoted in Madhursee Mukerjee, "Explaining Everything," *Scientific American* (January, 1996).

255   Says Brian Greene of Columbia University: Interview with Brian Greene, August 22, 2000.

256   "Well, he was smart enough . . .": Alice Steinbach, "Physicist Edward Witten, on the Trail of Universal Truth," *Baltimore Sun,* February 12, 1995, p. 1K.

257   At the age of twelve, Witten's letters: Jack Klaff, "Portrait: Is This the Cleverest Man in the World?" *The Guardian* (London), March 19, 1997, p. T6.

257   he has been involved with peace groups: Judy Siegel-Itzkovitch, "The Martian," *Jerusalem Post,* March 23, 1990.

258   Nathan Seiberg of Rutgers University: Mukerjee, "Explaining Everything."

258   strings are not really the fundamental particle: Hence the title of this chapter, taken from the title of talks given by M-theory pioneer Michael Duff of Texas A&M University.

259   Witten used to say that the M: Douglas M. Birch, "Universe's Blueprint Doesn't Come Easily," *Baltimore Sun,* January 9, 1998, p. 2A.

259   Lately, he has added: J. Madeline Nash, "Unfinished Symphony," *Time,* December 31, 1999, p. 83.

259   the physics of black holes: For a good discussion of black holes in M-theory, see Brian Greene, *The Elegant Universe* (New York: W. W. Norton & Co., 1999), chap. 13.

261   This could happen at the new Large Hadron Collider: "Discovering New Dimensions at LHC," *CERN Courier* (March 2000). Available on the web at http://www.cerncourier.com.

261   The other test will be a search: P. Weiss, "Hunting for Higher Dimensions," *Science News,* vol. 157, no. 8, February 19, 2000. Available on the web at http://www.sciencenews.org

261 they have so far studied the behavior of gravity: Researchers at Stanford University and the University of Colorado at Boulder are currently conducting experiments employing "desk-top" technology to test gravity at smaller distances.

261 He says, "I believe we have found . . .": John Schwarz, "Beyond Gauge Theories," unpublished preprint (hep-th/9807195), September 1, 1998, p. 2. From a talk presented on WIEN 98 in Santa Fe, New Mexico, June 1998.

ACKNOWLEDGMENTS

Thanks . . . to Alexei and Nicolai, for sacrificing their time with their dad for all the days it took for me to get this book done (though I know the loss is more mine than theirs); to Heather for being with them all the times I wasn't; to Susan Ginsberg for being the best agent in town, but most of all for believing in me; to my editor, Stephen Morrow, for recognizing and helping focus the vision, based only on the thinnest of proposals, and for gambling that I could (eventually) deliver, to Steve Arcella for his wonderful and caring work creating the illustrations; to Mark Hillery, Fred Rose, Matt Costello, and Marilyn Burns for their time, criticism, suggestions, and friendship, not necessarily in that order; to Brian Greene, Stanley Deser, Jerome Gauntlett, Bill Holly, Thordur Jonsson, Randy Rogel, Stephen Schnetzer, John Schwarz, Erhard Seiler, Alan Waldman, and Edward Witten for reading all or part of the manuscript; to Lauren Thomas for helping me translate some rather archaic French; to Geoffrey Chew, Stanley Deser, Jerome Gauntlett, Murray Gell-Mann, Brian Greene, John Schwarz, Helen Tuck, Gabriele Veneziano, and Edward Witten, for agreeing to be interviewed; and to the Minetta Tavern in Greenwich Village for providing an inviting meeting and writing place. Finally, I would like to acknowledge two other institutions: the New York Public Library, for having even the most obscure books on hand despite its under-funding; and Dover Publications, for reprinting, and thus saving, if not from obscurity, then at least from disappearance, many wonderful old books about physics, mathematics, and the history of science.

293

INDEX

Cartography, 43, 53, 54, 59
  medieval, 66
  spherical triangles in, 135
Catholic Church, 86–87
  under Charlemagne, 61
  Descartes and, 88, 92
  Galileo and, 86–87
  medieval, 66–68
Ceres, 195
Chanut, Pierre, 90–91
Charge, 241
Charged particles, 240–41
Charles V, king of France, 78
Charles the Great (Charlemagne),
  60–62
Chew, Geoffrey, 235, 239, 253
Christianity, 27. *See also* Catholic
  Church
Christina of Sweden, Queen,
  90–92
Church, Catholic. *See* Catholic
  Church
Church schools, 61, 62
Cicero (orator), 43, 45
Circle(s)
  Descartes' definition of, 34,
    81–82, 84
  Euclid's definition of, 81–82
  great, 134, 135, 139
Circumference of the earth, 41–42
City-states, Greek, 12
Cleopatra, 44
Clifford, William Kingdon,
  153–55
Clock, medieval, 66
Colleges, medieval, 64–65
Complementary pairs, 225
Confucius, 11
Congruence, 123

Conic Sections (Apollonius), 47
Coordinate geometry, 53
Coordinates, Cartesian, 83
Copernicus, 78
Coupling constants, 241–42
Critique of Pure Reason (Kant),
  117–18
Crouch, Henry, 210
Crusades, 62
Curry, Paul, 32, 33
Curved space, x–xi, 96, 105
  Clifford on, 153–55
  Gauss on, 128–29
  general relativity and, 207–8
  impact on mathematics, 146–
    147
Curves, equations for, 84
Cyril, 47, 48

Damascius, 46–47
Dark Ages, 49, 62
Dedekind, Richard, 71
Delta function, 72
Descartes, René, x, 49, 79–89
  Catholic Church and, 88, 92
  definition of circle (or ellipse),
    81–82, 84
  Discourse on Method, 88
  distance formula, 86
  on Greek geometry, 81
  on lines, 83–84
  in Sweden, 90–92
*Dialogue on the Two Chief
  Systems* (Galileo), 87
Differential geometry, 7, 128,
  138–42, 143, 205
Diogenes Laertius, 12
Dirac, Paul, 72, 222, 240
Dirichlet, Johann, 141

fundamental constants sought
by, 68
gravity and, 251–52
kinds of, 255, 258
relativity and, 249
space and, 245–48
spinning, 250
standard model and, 242
superstrings and, 252, 254
troubles with, 249–54
vibration of strings, 244–46,
247
Strominger, Andrew, 255, 260
Strong force, 237, 240, 241
Sturm und Drang, 112
Supergravity, 252
Superstrings, 252, 254
Supersymmetric particles, 261
Supersymmetry, 250, 252
Susskind, Leonard, 237
Sybaris, 27
Syene, town of, 41
Symposium, Greek, 13

Table-rapping, 116
Tachyons, 249
Taurinus, F. A., 116, 121
Taxes in ancient Egypt, 6
Teller, Edward, 213
Telys, 27
Tentamen, 119
Tesla, Nikola, 211
Thabit ibn Qurrah, 102–3, 144
Thales of Miletus, 11, 12, 13–16
on physical space, 14–15
Pythagoras and, 15–16
systematization of geometry by,
14
Theon, 46

*Théorie des nombres (Theory of
Numbers)* (Legendre), 136
Thomson, Sir William (Lord
Kelvin), 172
Time
absolute, 190
gravity and, 201–4
Kant's view of, 177
local, 174–75
medieval concept of, 65–66
in M-theory, 258
in Newtonian mechanics, 155
Newton's view of, 155–56
proper, 188
relativity of, 187–88
space and, 175
universal, 174–75
Tomanaga, Sin-itiro, 241
Topographia Christiana
(Boethius), 46, 49
Topographical maps, 73
Topology, 246
Transverse vibration, 245–46
*Treatise on Electricity and
Magnetism, A* (Maxwell),
163
Triangles
similar, 104–5, 121
spherical, 129–35
Triangular numbers, properties of,
18, 19
Tufte, Edward, 74
Turner, Peter, 104

Uncertainty principle, 223–27
Undefined terms, need for,
143–45
Unified field theory, 213, 228, 231
Uniform motion, 195–98